Capturing the Public Value of Heritage

THE PROCEEDINGS OF THE LONDON CONFERENCE

25–26 January 2006

Fig 1 In 1986 Stonehenge was among the first places in the UK to be inscribed as a World Heritage Site – in the language of UNESCO, a place of Outstanding Universal Value. Since then the future of this iconic monument, severed from its surrounding prehistoric landscape by two busy roads and serviced by an ugly and intrusive visitor centre, has been the subject of endless debate. Now, 20 years on, the concept of public value – what the public values – may at last be one of the keys to a better future for England's most famous archaeological site. © English Heritage

Capturing the Public Value of Heritage

THE PROCEEDINGS OF THE LONDON CONFERENCE
25–26 January 2006

Edited by Kate Clark

Acknowledgements

The conference was organised jointly by Kate Clark (Heritage Lottery Fund), Frances MacLeod (Department for Culture, Media and Sport), Duncan McCallum (English Heritage) and Tony Burton (National Trust). The event was managed by Jane Callaghan and her team from Multimedia Ventures and was held at the Royal Geographical Society in London. We owe a real debt of gratitude to Jane and her team, and to the staff of the RGS for making the event such a success. We would like to thank each of the speakers and in particular our guests – David Throsby from Australia, Christina Cameron from Canada and Randall Mason from the USA. Special thanks also go to the members of the Citizens' Juries – Bunney Hayes, Dolly Tank, Michael Rosser and Leila Seton. The stallholders made a vital contribution to the two days, and we would like to thank the Castleford Heritage Trust, the Department for Culture, Media and Sport, English Heritage, the Heritage Lottery Fund, the Local Heritage Initiative, Market Leader Research, Museums Libraries and Archives and the Royal Society for the Protection of Birds. We were delighted that the children of Castleford and the members of Urban Roots were able to be there to inspire us. Many other people were involved in making the event a success, and we would like to thank Ed Badger, Karen Brookfield, Judy Cligman, Jamie Cowling, Geoff Dawe, Joanna Finn, Deborah Lamb, Helen Monger, Deborah Owens, Katie Roberts, Robert Smith, Catherine Ware, Corinna Woodall, Jon Williams and his team, Shane Winser and the members of UKHERG. Our special thanks go to Rowan and Susan Whimster of Whimster Associates for co-ordinating this volume and liaising with the contributors and designer, and to the speakers for their kind co-operation in contributing to it.

Published by
English Heritage
Kemble Drive
Swindon SN2 2GZ
www.english-heritage.org.uk

on behalf of the conference sponsors:
The Department for Culture, Media and Sport, English Heritage,
the Heritage Lottery Fund and the National Trust

Copyright © 2006 the individual authors or their employing institutions

First published May 2006

ISBN 1 905624 10 7
ISBN 978 1 905624 10 2

Product Code 51216

British Library Cataloguing-in-Publication Data
A CIP catalogue record for this book is available from the British Library.

All rights reserved
No part of this publication may be reproduced or transmitted in any form
or by any means, electronic or mechanical, including photocopying,
recording or any information storage or retrieval system,
without permission in writing from the publisher.

Brought to publication by Whimster Associates Ltd
Designed by Libanus Press Ltd
Printed in England by Hawthornes Ltd

Contents

Introduction 1

1 The Public Values of Heritage

From consultation to conversation: the challenge of *Better Places to Live* 7
  Rt Hon Tessa Jowell, MP
Public value as a framework for analysing the value of heritage: the ideas 14
  Robert Hewison and John Holden
Capturing the public value of heritage: looking beyond the numbers 19
  Accenture
Heritage, democracy and public value 23
  Ricardo Blaug, Louise Horner and Rohit Lekhi
Discussion 1. Public value: how useful is it as an idea? 28

2 The Instrumental Benefits of Heritage

Sustainable communities: the economic, social and environmental benefits of heritage 33
  Baroness Andrews
The value of cultural heritage: what can economics tell us? 40
  David Throsby
Capturing the impact of museums on learning 44
  Sue Wilkinson
Public space: public value 48
  Julia Thrift
Heritage on the front line: the role of a heritage champion in North Yorkshire 53
  Heather Garnett
From significance to sustainability 59
  Kate Clark
Discussion 2. The instrumental benefits of heritage: how are they measured? 61

3 The Intrinsic Values of Heritage

Community, identity and heritage 65
  David Lammy, MP
Capturing the value of places 70
  Sir Neil Cossons
Value and integrity in cultural and natural heritage – from Parks Canada to World Heritage 71
  Christina Cameron
Why do places matter? The new English Heritage *Conservation Principles* 79
  Edward Impey
Capturing the opinions of people 85
  Dame Liz Forgan
The value of heritage: what does the public think? 86
  Deborah Mattinson
The Citizens' Jury 92

4 The Values of Heritage Institutions
    Conference overview 93
        Kate Clark
    Discussion 3. Whose values matter? 95

Conclusion 99

Biographies of Speakers 100

References 102

Delegates Attending the Conference 103

Introduction

'What is aught, but as 'tis valued?'
Shakespeare, *Troilus and Cressida*

In March 2005, Tessa Jowell, the Secretary of State at the Department for Culture, Media and Sport, produced an essay challenging the sector to find a new language to describe the importance of the historic environment (*Better Places to Live*, Jowell 2005). She put a number of challenges to the sector, including the need to increase diversity in both audiences and the workforce, to capture and present evidence of the value of heritage, to contribute to the national debate on identity and Britishness, to 'create public engagement and to widen the sense of ownership of the historic and built environment'.

Meanwhile, the Heritage Lottery Fund, the major funding organisation for the heritage, had undertaken a series of Citizens' Juries, exploring public views on heritage and, in particular, a series of projects funded by the Heritage Lottery Fund. The Fund had also commissioned a review of the evidence for the impact and benefit of the £3billion they had given to approximately 15,000 projects. Demos – who undertook the work – had suggested public value as an organising framework.

English Heritage, the government's adviser on the historic environment, then produced the first draft of a new philosophy to underpin decisions about the historic environment. The document put the concept of value or significance at the heart of decision-making, and also called for heritage professionals to operate in a new and more transparent way (English Heritage 2006).

And finally, the National Trust, the largest voluntary organisation in the UK – conscious of the need to produce more robust evidence of impact – had been working with the consulting firm Accenture in order to find new ways of measuring performance and heritage sites that were open to the public.

The question of value and how to measure it in public sector organisations was also under debate in other arenas. The charter for the BBC was due to expire and the Corporation had pioneered the use of public value as a framework within which to set their bid for renewal; the Department for Culture, Media and Sport and others had asked the Work Foundation to explore the application of ideas of public value to arts and culture while Gavin Kelly, Geoff Mulgan and Stephen Muers had produced a paper for the Strategy Unit of the Cabinet Office, setting out the idea of public value as an analytical framework for public service reform (Kelly *et al* 2002). All of this was, of course, building on the work of Mark Moore, author of *Creating Public Value* (1995).

Early 2006 was therefore an opportune time to bring together people from the heritage community and beyond in order to look at the concept of public value, and how it might be applied to thinking about heritage. We had anticipated something small and fairly academic, perhaps of interest to a select few heritage policy-makers; in the event, more than 400 people from heritage and beyond spent two extraordinary days at the Royal Geographical Society listening to everyone from a young urban hip-hop group to three government ministers. Many of us came away inspired to think again about heritage and what it means to people.

The concept of 'public value'

At its most basic, 'public value' is simply an analytical framework that can be used to look at how public sector organisations operate. It starts from the premise that such organisations are there to add or create value for the public, and that therefore the best way of measuring their success is to look at it in terms of what the public cares about. Public value is the equivalent of shareholder value in the private sector, but it is expressed through the democratic realm, rather than the market place.

As Moore has explained, private sector organisations create value by making money for shareholders and providing customers with goods and services. For public sector organisations, their shareholders are in effect the legislators who provide resources and authorise them to operate, and their customers the people who benefit from (but do not necessarily pay directly for) their services. Public sector organisations need to create value upstream for those who provide resources and downstream for the people who use their services, but that value is not necessarily seen in terms of profit.

For Moore, public managers must consider three things – the public value that their organisations seek to produce (which can be seen in its aims and objectives), the sources of legitimacy and support they can rely on to provide authorisation and resources, and the operational capacity including the staff, financial and technical resources. Each of these has to fit together; it is not enough to consider only one.

Public value is in part a reaction to the application of what is known as the 'new public management' to public sector organisations in the 1980s and 1990s. This involved adapting private sector business-management thinking to public sector organisations. And while some of that has been beneficial – for example, more transparency and better management structures, as Kelly et al note (2002, 9) – there have also been disadvantages. New public management also tended to emphasise narrow ideas of cost-efficiency – focusing on how things were done rather than whether they were the right things (for example, counting 'finished consultant episodes' in hospitals). As they state, 'Those things that were easy to measure tended to become objectives and those that count were downplayed or ignored.'

Ultimately, public value challenges the conventional 'market failure' rationale for government action. Instead of simply providing goods and services that the market does not, public value suggests that people have a more subjective, values-based approach to what they want from government.

Public value is often criticised as being based simply on a crude understanding of what the public wants; it is important that in applying it, service-providers are responsible to what is valued but do not just pander to ill-informed preferences. While it certainly creates a greater role for citizens in helping to shape what an organisation does, this is something that needs to be mediated or refined. The process of engagement is not just about collecting data but deliberation and education. And ultimately there are checks in place through the press and the democratic process that should prevent leaders from seeking to impose views that cut against the grain of popular opinion (Kelly et al 2002, 7).

Mark and Gaylen Moore have worked with a group of State Arts Agencies in the USA, in order to see whether public value provides a useful framework to help them understand strategic choices (Moore and Moore 2005). The resulting report should be of interest to anyone in the cultural sector – the question, however, is whether it works as well for heritage.

Heritage and public value

In fact, there are special reasons for thinking about heritage in the light of public value. At its most basic, heritage is what people value and want to hand on to the future. Heritage is very broad – it can cover everything from land and biodiversity, to buildings and landscapes, collections and even intangible heritage such as language and memory. In fact what makes something part of our heritage is not whether it is a building or landscape, but the value that we place on it.

Value therefore remains at the centre of all heritage practice; it is what justifies legal protection, funding or regulation; it is what inspires people to get involved with heritage. Indeed, in public value terms, something is only of value if citizens – either individually or collectively – are willing to give something up in return for it (Kelly *et al* 2002, 4). This happens for heritage when a developer faces constraints over new development in a historic building, lottery players' money is used to fund a heritage project or a volunteer gives up their time.

But of course there are different kinds of values in heritage.

There are the values that we put on something that mean that we want to keep it – perhaps after its useful life. These may be aesthetic, social, scientific or historical; we may value something for the story it has to tell about the past, or because it was associated with events or people. Something may have an intense personal value, or it may hold memories for us as individuals or as communities. The stewardship role of heritage organisations is about looking after those assets that people value, whether through protecting buildings, funding projects or opening sites to the public.

The values people put on heritage may – or may not – be different from the benefits that people gain from a heritage activity, such as conserving a building or volunteering. Those benefits can range from individual benefits – such as greater confidence or new skills – to community, indeed nationwide, benefits – prosperity, a sense of identity, cohesion. A public park, for example, might be protected as a fine example of historic landscape design; the benefits that arise from restoring that park might be very different – it could become a safer and more welcoming place to take exercise or meet friends.

This distinction between instrumental and intrinsic benefits was identified for the arts by the RAND Corporation. It also came out very strongly in a recent project run by the Getty Conservation Institute that analysed value at four major heritage sites across the world (de la Torre *et al* 2005). While the distinction may not always be very clear in, for example, creative arts, for heritage there can often be very real differences between the so-called 'intrinsic' values that we ascribe to a place or object, and the instrumental benefits that arise from funding or conserving it.

The third type of values are institutional values. The public value framework stresses the importance of how organisations behave, and the need to generate trust and legitimacy. As heritage involves looking after something on behalf of the public, such issues would seem to be as – or more – important to heritage as to any other sector.

Because heritage bridges both culture and environment, there are potentially many different ways of capturing its impact and value. From an environmental perspective, heritage can contribute to many of the goals and objectives of conservation and sustainable development, and indeed thinking about sustainable development can be a helpful way of framing heritage issues. Alternatively, from a cultural perspective, heritage can make a contribution to social outcomes such as creativity and skills for individuals and for communities.

The advantage of the public value approach over other frameworks is that it places concepts of value at the centre of thinking and it is that – more than anything else – that makes public value such an attractive proposition. But like all theoretical frameworks, the critical factor is whether it can work in practice.

Capturing the Public Value of Heritage was an opportunity to begin to find out whether public value was a practical tool.

The conference

When we put together this conference, we started with the three types of value that heritage organisations need to think about – the 'intrinsic' value of the heritage, the 'instrumental benefits' that arise from caring for it and the 'institutional' values that heritage organisations need to think about.

We tried to bring together a range of people to discuss this from different perspectives. We heard from representatives from the Heritage Lottery Fund Citizens' Juries; people on the front line in local authorities such as Cllr Heather Garnett, a Local Authority Heritage Champion; and academics such as David Throsby, who has been wrestling with fundamental ideas about value. We tried to represent a range of heritage from buildings and collections to parks and biodiversity, but inevitably we could not touch on all of it. And young people from Castleford and South London performed at the event, reminding us of the sheer delight that can come from engaging with heritage.

This volume brings together the results of that conference. Our aim has been to capture as quickly as possible what happened and what was said, rather than spend time creating polished pieces. We are immensely grateful to all of the speakers for providing their texts so quickly and for being prepared to contribute to this volume. We hope we have encapsulated the spirit of the event, to enable those who were not there to share it and to make a contribution to a debate which we hope will have moved on, even as this goes to press.

Kate Clark

1 The Public Values of Heritage

Rt Hon Tessa Jowell, MP
Secretary of State for Culture, Media and Sport
From consultation to conversation:
the challenge of *Better Places to Live*

Robert Hewison and John Holden
Demos
Public value as a framework for analysing the value of heritage:
the ideas

Accenture
Capturing the public value of heritage:
looking beyond the numbers

Ricardo Blaug, Louise Horner and Rohit Lekhi
The Work Foundation, in association with the Research Republic
Heritage, democracy and public value

Nick Higham (chair)
Arts and media correspondent, BBC
Discussion 1
Public value: how useful is it as an idea?

From consultation to conversation: the challenge of *Better Places to Live*

Rt Hon Tessa Jowell, MP
Secretary of State for Culture, Media and Sport

Introduction

I would like to start this incredibly important conference by signalling a health warning: beware of public value 'blah', that is to say discussion that lacks definition, intellectual rigour and substance. The idea of public value – which I believe in very strongly – will only survive if we are rigorous in its definition and application.

Mark Moore, the inventor of the public value concept, said that 'Public value is what the public values'. There is a profound truth behind that simple definition, and it takes a large and distinguished gathering like the one here today to begin to unpack that truth and to apply it to the work we do together to sustain, nurture, enrich and proselytise the heritage offering.

Thanks to the work of English Heritage, the Heritage Lottery Fund and the National Trust – among others – we have increasingly come to understand just how valuable our heritage is, and I would like to congratulate and thank them for joining with officials in my department in putting on this conference and bringing us all together.

Value of culture

We have come a long way in the two years since I first published my essay on the *Value of Culture* (Jowell 2004). I wanted to spark a debate about the way culture influences our lives. And I think to quite a fair extent, I succeeded. Plenty of you agreed with the ideas I put forward in that essay. Plenty more of you challenged me for not putting enough emphasis on heritage – and you were right.

That is why last year, I published *Better Places to Live* (Jowell 2005). Again, the point was to try and move the debate on, and highlight the unique role that heritage can play to build bridges between our past and our future.

Best of Britishness

Today, the parameters of our conversation are shifting once again. For instance, lots of people are talking about Britishness, and a little later this year, I will be publishing another essay that will look at the subject in more detail. But now, I want to explore the links

Fig 2 Children role playing in the Great Chamber at Sutton House in Hackney. They came from a school in Tower Hamlets and their visit was part of their core curriculum history studies. Learning is one of the key instrumental benefits of heritage – using historic places can open people's eyes to their shared identity and link the past with the future. © NTPL/Chris King

Fig 3 The Durbar Room at Osborne House – a celebration of the cultural brilliance of the Indian sub-continent and its contribution to British society during the reign of Queen Victoria.
James O Davies © English Heritage

between public value and heritage and the public benefit that comes from developing a sense of shared identity.

In a way, I see our built heritage as being a more permanent expression of all the people who have come before us. Modern Britain is a country of many peoples, many stories, and many cultural perspectives and experiences, but it is vital that we understand that this is not a particularly modern phenomenon (Fig 3). Historically, the British Isles has always been a magnet for many different people, from many different backgrounds. Britain itself was forged in the furnace of diversity, bringing Scotland, England and Wales together into one whole. So our history is one of difference and broadly, of tolerance. There is nothing new there.

By providing a tangible link with our past, our physical heritage is helping to underscore the fact that Britishness does not mean coping with difference – Britishness is dependent on difference.

Public realm

That is why I have always regarded our historic environment as being a vital part of the public realm – part of what I would define as those shared spaces and places that we hold in common and where we meet as equal citizens. The places that people instinctively recognise and value as not just being part of the landscape or townscape, but as actually being part of their own personal identity.

That is the essential reason why people value heritage. As I said a few weeks back, at the launch of this year's *Heritage Counts* report (English Heritage 2005), all of us need to have a richer and fuller understanding of this, and how it should translate into decisions about public policy, public spending and public management. We also need to understand that there is a symbiotic relationship between identity and the public realm: they are both

Fig 4 The British Museum, looking north from the South Portico into the Great Court. © Nigel Young

underpinned by a commitment to treating people as having equal worth; they both exemplify plurality; and they both make the point that our success as a nation has in no small part been based on the fact that we have been willing to embrace new people, new ideas and new influences. Individual differences have never diminished the concept of Britishness; in fact, I would go as far as to say that our differences have always enhanced us, as a nation, and helped to make us who we are. In fact we define ourselves increasingly in terms of our common values – fairness, openness and tolerance.

But I am also acutely aware that this is very sensitive territory for politicians to explore. My colleagues and I have a legitimate role in the debate about Britishness, but it is just one among many roles and it must be a limited role. Many people are allergic to hearing politicians, or anybody else, trying to be too prescriptive about Britishness, and seeking to prescribe the form of identity that an intensely private set of views and experiences can shape is a key problem in this debate.

I do believe that there is a robust content to Britishness; I do believe that British values and ideals can be asserted and promulgated, not least through some of our great national institutions like the BBC and the British Museum (Fig 4) and many others that I am privileged to sponsor; and I do believe that political leaders on the centre-left should never let the right monopolise the debate about national identity.

But I also believe that for most people, Britishness matters not because of its national significance but because of its personal importance. We sometimes recoil when we hear other people try to describe it or analyse or encode it because in so doing we know that we risk the intimacy of what it means to us in our own private world. Our sense of Britishness goes well beyond national symbols and institutions. In fact, much of the debate begins by looking down the wrong end of the telescope – it starts with the individual, not the state. I think that for most people, Britishness is just one aspect of a much richer and more subtle sense of identity – it is about how we feel; it is about who we think we are; it is about our sense of belonging. And this will be different for each person.

So people who are interested in this issue should unite with politicians to show some humility, and focus on what we can do best, which is to ensure that those places and spaces exist where people have a chance to learn, to reflect and to build their own sense of who they are, how they fit in and what they share with others. In a sense, it is the job of politicians and government to provide this infrastructure, the hardwiring if you like, where our identity is played out. Of course, the heritage sector has a huge part to play in creating those places and spaces where that can happen.

Olympics

These opportunities to reflect and to celebrate were certainly in evidence last summer. We saw it in the wake of the terrorist bombings on 7 July last year, and it was this spirit of inclusive British cultural identity that won us the right to host the Olympics on 6 July (Fig 5). We also set a precedent of displaying the value of tolerance despite our sense of outrage that terrorists could threaten our way of life.

The year 2012 is going to be a huge moment of national realisation and national consciousness, and we have got to make the most of it. It is also going to generate private memories, just as some people still cherish memories of when Britain last hosted the Olympics in 1948. I know work is already under way, for example developing a 'Heritage Trail' for the Olympics, but there is so much more we can do.

Fig 5 On 6 July 2006 huge crowds converged on Trafalgar Square, the historic heart of the nation's capital, to celebrate Britain's winning bid for the 2012 Olympics. © Hayley Madden

The public value framework

Later in this conference, Demos and the Work Foundation will share the latest thinking about how we can use the concept of public value as a framework, to find new ways to engage the public and deliver services and make sure that we *are* making the right links.

As part of this work, Demos identified the 'three-legged stool' approach to public value, and this is an approach which the heritage sector has already adopted. Indeed, this conference has been designed around these three key concepts of measuring the intrinsic value, instrumental value and institutional value of our heritage.

I know that Simon Thurley likes to make a distinction between the 'established value' of heritage and other forms of value. 'Established value' is something on which everybody agrees and which is embodied in the work of the listings and scheduling systems, but there are other, more contested aspects of heritage value such as aesthetic value, community value, evidential value and historical value. We need to explore all of these and maybe to add to them in order to see how they might be part of a public value framework for heritage. There is still a debate to be had, because while theories of new public management have been pretty successful at getting the public sector to focus on delivery, some really challenging issues remain.

Management consultants like to chant that 'what gets measured, gets done'. The trouble, however, is that you cannot measure everything of value and simply adopting output targets risks distorting effort. Then, there is the delivery paradox. Although our public services are objectively improving, in too many cases the public cannot or will not believe the evidence. Lastly, while most people would accept that cost-benefit analysis is a pretty good way of trying to decide where to spend public money, it is pretty tough to measure all the value, or benefit, that we can get from certain goods.

Work Foundation project

In order to better this, my department is working closely with the Work Foundation project to try and find practical ways to capture what people value, and to translate that information into decisions about priorities for public policy, public spending and management. But before we do, we will need to find answers to a lot of questions, not least because as an emerging concept, public value has not yet been properly defined. We will also have to try and clarify the murky concept of what it means to be a citizen – much, much more than being just a consumer.

This touches on my particular interest in the public value process, namely how it can help us to understand how our public institutions can create and embody value for citizens. The market place can tell us how many people visited a particular museum or how much profit a particular show or event made, for example, but when it comes to putting a value on things like trust, fairness and accountability, it has failed miserably.

Asking the public

Adopting the public value approach would be a radically different way of doing things and for it to be successful, it will require a radically different mindset to the one that many of us have today. It means taking a genuine interest in what our citizens think, and not just consulting in a ritualistic and formulaic way because we have to. It means engaging a much wider swathe of society, particularly people who are socially excluded and people from ethnic minority communities, and not just talking to the usual suspects. And it means adopting an approach where we do not just care that something is delivered; we also care about its quality, and how it was delivered.

The BBC experience of public value has helped it begin to transform the way it works. It has also underscored the fact that talking to the public – the people who pay for it and on whose behalf the BBC Trust will hold the BBC to account – is not just a nice idea or optional extra. It should be the heartbeat of the decision-making process for public institutions. This has implications for the Department for Culture, Media and Sport, and also for our partners, including the heritage sector.

Challenges facing the heritage sector

In a public value world, our public service agreements would be less about what we want to deliver, and more about what we know the public want from us. As well as targets for attendance, we would also need targets for satisfaction and quality. In turn, that would lead to significant changes in what we were choosing to fund, and how we were choosing to do it.

A public value world would include a lot more 'co-production of services' at the local level. Instead of funding what we think is important, we would start by asking people what is important to them and then consider how to protect it. In terms of heritage, that would mean asking the public which buildings and open spaces they value in their local area, and then allocating funding accordingly. Instead of experts making all the decisions, experts would share their knowledge with the public, and facilitate people making more of their own informed judgements. The whole process of consultation and collaboration is becoming less crude and more informed, and so the art of promoting greater involvement is becoming better understood.

Am I describing a radical departure from the current way of doing things? Absolutely. But it is the only way to maintain our legitimacy, and ensure that our priorities are shared by the public. If we do not adopt a more public value way of doing things, then we cannot just take it for granted that we will continue to get the support we deserve.

Responses to *Better Places to Live*

We all need to make the quality of our engagement with the public a priority, and make sure that they really are involved in the decision-making process. That is why I asked a series of questions at the end of *Better Places to Live*, which attracted a fair bit of attention. I have published a summary of the responses and the key points, which I think make for some interesting reading (DCMS 2006).

You told me that the debate needed to include all parts of the public realm, and not just our urban heritage or the big iconic visitor attractions. You also welcomed the repositioning of the debate onto the value and the contribution that heritage makes to our quality of life, our identity and our sense of pride and national self-esteem. You also recognised that we need to address the lack of diversity in the heritage workforce, and pointed to the Museums Association's Diversify programme, as an example of how it could be done.

We did not agree on everything, of course. You did not much like my idea of using digital recordings of buildings as an alternative where the buildings cannot be kept intact. Instead, you argued for increased public funding to ensure that we can conserve as much of our heritage as possible for future generations. Well, ideally I would like to do both, and I certainly agree with you that digital recordings should never become a cheap substitute for the bolder alternative of preserving valuable parts of the physical fabric of our history.

If you are looking for a testing ground for public value, you would be hard pressed to find a better subject. We all know that local communities will rally round when local buildings are being threatened by demolition, and it is fascinating that the debate sparked by the Heritage Protection Review has led to an 86 per cent increase in requests for listing. But I wonder whether the buildings that communities value are always the same as the ones the experts want to save.

One thing where there is 100 per cent agreement between us is that all of these decisions need to be informed decisions. That is why it is so important that organisations such as Save continue to campaign, and to engage the public in a debate about the issues. Of course, we cannot save every historic building – and I do not think we should. But where we are demolishing our historic assets, we need to be absolutely sure that it is the right thing to do – and that we are doing it for the right economic, social and environmental reasons.

The public's refined preferences

The Work Foundation calls this way of working 'using the public's refined preferences'. It means replacing consultation with conversation, and dramatically changing the role of experts – using experts to inform and empower the public, rather than simply cutting the public out and leaving everything to the experts. Refined preferences also poses some interesting questions vis-à-vis funding. In an increasingly wealthy society, there is also a question as to the proper balance between public funding philanthropy and other forms of funding. This builds on the idea that what we hold, we hold in trust for the public.

We cannot just assume that we know what the public wants as a priority for public investment – we have to start asking them.

Heritage Lottery Fund

This may sound like a moderately radical concept, but the public value approach is already informing more of our work than you might think. The Heritage Lottery Fund, for example, has already started this discussion through its work with Citizens' Juries. By doing so, it is giving a voice to communities that before now have been silent. For the first time, our young people are talking about their heritage; our veterans are recording their experiences of the war; and families who many generations ago were affected by the Slave Trade are sharing their experiences as 2007 approaches.

Last year, four out of five of the fund's awards were for less than £50,000, many going to small community groups. And, of course, it is continuing to fund large-scale projects too. Today, I can announce that the National Heritage Memorial Fund Trustees have awarded a further £15 million to large projects, including £7 million for The People's History Museum in Manchester (Fig 6). This unique museum is dedicated to the 'extraordinary story of ordinary people', and will offer a new look at the lives of working people over the last 200 years. It is one of my very favourite places and it is a great example of how heritage, culture and Britishness can and should intersect.

Fig 6 At the conference, Tessa Jowell announced that the Heritage Lottery Fund had awarded £7.18 million to The People's History Museum in Manchester. This will allow the museum to consolidate its current two sites into one, providing much-improved visitor facilities and increased access to the museum's collections. © People's History Museum

Conclusion

So today, as I open this conference – which is an enormous privilege – the heritage sector has a real opportunity to put itself to the public value test. Passing that test will not just mean that the sector can continue to protect and enhance our heritage; it will also ensure that the sector drives itself forward to be a relevant, valued and vitally important part of a successful, diverse and modern Britain.

Public value as a framework for analysing the value of heritage: the ideas

Robert Hewison and John Holden
Demos

JH: Since this a conference about heritage, let's start with a little history. We're on the platform today because of a piece of work that we did for the Heritage Lottery Fund two years ago that resulted in a report called *Challenge and Change* (Demos 2004).

RH: I'm challenge – he's change.

JH: Before that work began, and independently of the project, we had recognised the general dissatisfaction that was being felt throughout the cultural sector because culture was being accounted for using inappropriate and inadequate systems of measurement. Fundamentally, people were being asked to measure the wrong things in the wrong way, a lot of the data that was produced was spurious, and it wasn't being usefully applied anyway.

RH: That was the clear message of a conference jointly organised by Demos and AEA Associates, the National Gallery and the National Theatre in June 2003, called *Valuing Culture*. Tessa Jowell was at that conference, and it was partly in response to it that she produced her personal essay, *Government and the Value of Culture*, the following year. In the meantime Demos has continued to engage with the issues of measurement, accountability and above all value, including through the work it has done for the Heritage Lottery Fund.

JH: But let's return to *Challenge and Change*. As far as the Heritage Lottery Fund was concerned, in 2003 imminent legislation on the Lottery threatened to have a significant impact on the way it would operate in the future – indeed, like other Lottery distributors, at that stage, it did not know if it had a future at all beyond 2009. On top of that, the Department for Culture, Media and Sport had started a review of the entire heritage sector, with mergers or a redistribution of responsibilities as an option.

RH: Our job was to look at the mass of research data that the Heritage Lottery Fund had accumulated over the nearly ten years of its existence and from that produce some kind of understanding of how the Heritage Lottery Fund had been meeting the social and economic expectations placed upon it by government.

JH: In order to do that we developed a conceptual framework that enabled us to talk not just about outputs and outcomes, but also about the range of values that drives the heritage sector. This took us beyond the reductive economistic language that has become all too familiar to cultural organisations in recent years.

RH: What was particularly interesting about the Heritage Lottery Fund was the way that it had shifted the idea of the value and importance of heritage away from being something that is exclusively determined by experts on behalf of society, to one that recognises the importance of widespread public participation in identifying and caring for what is valued collectively. The work of the Heritage Lottery Fund had in fact broadened the social base for the enjoyment of heritage so that there is now an acknowledged diversity of contributions to the national story.

JH: What we want to do this morning is to set out the conceptual framework that we developed, because it explains the types of value that are integral to the concept of heritage and the context in which those values are articulated. What we are talking about has become known as 'cultural value'.

RH: Schematically, heritage generates three types of cultural value. The first is the value of heritage in itself, its intrinsic value in terms of the individual's experience of heritage intellectually, emotionally and spiritually. It is these values that people refer to when they say things like 'This tells me who I am', or 'This moves me' or quite simply 'This is beautiful'. Of course, people will differ in their individual judgements, and because these values are experienced at the level of the individual, they are hard to quantify – yet we all know they exist. To quote Tessa Jowell in her essay on heritage, *Better Places To Live*: 'Historic sites, objects, modern or historic architecture can move us in just the same way as literature, music and the fine arts.' (Jowell 2005.) But *how* they move us, and how far, is not yet part of the calculus of funding or service level agreements.

JH: The second type of value can be termed 'instrumental': this refers to those ancillary effects of heritage where it is used to achieve a social or economic purpose. Urban regeneration is one obvious example, but there are also less clearly connected objectives, such as the reduction of crime. Instrumental values are generally expressed in figures, but as we have already pointed out, the measurement of such benefits – social or economic – is highly problematic.

RH: The third value is what we call 'institutional value'. This relates to the processes and techniques that organisations adopt in how they work to create value for the public. Institutional value is generated, or destroyed, by *how* organisations engage with their publics; it flows from their working practices and attitudes and is rooted in notions of the public good. Through its concern for the public an institution can achieve such public goods as creating trust and mutual respect between citizens, enhancing the public realm and providing a context for sociability and the enjoyment of shared experiences. Heritage organisations should be considered not just as repositories of objects, or sites of experience or ways of generating cultural meaning, but as creators of value in their own right.

JH: These three categories of value – the intrinsic, the instrumental and the institutional – can be visualised as forming the three angles of a triangle (Fig 7). This is an equilateral triangle; the equal angles are there to suggest the equal importance of the intrinsic, the instrumental and the institutional.

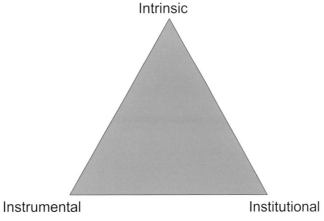

Fig 7 The Demos triangle of heritage values

RH: The point about cultural values is that they are plural, and their relative importance will depend upon your individual perspective. The question now is: which of these three sets of value is important to whom? Again, very schematically, we suggest that there are three distinctive groups of people with an interest (Fig 8).

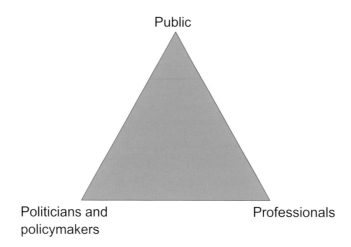

Fig 8 The triangle of heritage stakeholders

JH: Another triangle, and not a pyramid, even if it looks as though one particular group has come out on top. First, there are the politicians and policymakers, who for more than thirty years have identified most strongly with instrumental values. Economic benefits under Thatcher, social outcomes under Blair.

Then there are the professionals, the people whose work, in this case, it is to administer and care for the heritage. Their sense of vocation is driven by intrinsic values, while their professionalism is a vital part of institutional value.

Finally, there is the public. This is obviously not a homogenous group, but while they will have individual and distinctive preferences, what they value is primarily what we have characterised as intrinsic values. They are also very interested in how they are treated – in other words they are acute judges of institutional value. Economic regeneration and social inclusion are not the first thoughts of visitors to a heritage site. What *they* are looking for – apart from the lavatories and the shop – is an imaginative engagement, a sense of place, the satisfaction of curiosity and the feeling that they have gained from the experience.

RH: The conceptual model that we have proposed is, of course, applicable not just to the heritage sector, but also right across the field of publicly funded cultural activity. The specific application of the methodology to any one part of the field will reveal that in detail there is a different balance between the sets of value produced, depending on the mission, scale and structure of the particular organisation or organisations under review.

JH: In the case of heritage, in our report *Challenge and Change*, we identified a set of public goods that were associated with the Heritage Lottery Fund's strategic objectives. Some might be described as generic, and can be found throughout the cultural field, others were specific to the heritage sector, or to the organisation itself. These were:

- Stewardship: preserving the past – the intrinsic value of heritage – in the interest of the future
- Trust: that is, producing enhanced trust in public institutions
- The promotion of equity and fairness in the distribution of Lottery money
- Efficiency and resilience in organisations that receive funding
- Value for money in terms of the costs of delivery

- An enhanced sense of well-being and an improved quality of life for visitors
- The contribution that heritage makes towards general prosperity and employment
- The encouragement of learning and personal development
- The strengthening of communities, particularly by rediscovering a sense of connection to place.

RH: This demonstrates the rich mix of values that heritage creates: some are intrinsic, some instrumental and some institutional. They combine to create a structure where values reinforce each other – and we believe that the interests that we have identified are all represented and served within it. It also shows that a variety of methodologies will be needed to identify, express and communicate the particular manifestations of cultural value that each aspect of the general public good represents. For example, value for money can be calculated by making number-based comparisons between different organisations, whereas the emotional attachment to place cannot. You will be hearing from others this morning on what are the most appropriate ways of measuring some of these things; our point here is simply to argue that the concept of cultural value creates a context in which these different measures can reinforce rather than contradict each other.

JH: All three subsets of cultural value – the intrinsic, the instrumental and the institutional – are valid and important. We are not saying that instrumental values are irrelevant, or that heritage does not produce instrumental benefits. Neither are we trying to engineer a return to 'art for art's sake'. What we do argue is that the language of cultural value gives the sector an opportunity to renegotiate the relationship between the three interest groups; politicians, professionals and the public. This is necessary because in the past the discourse of heritage has become almost exclusively a conversation – even an argument – between the professionals and the politicians, and one overwhelmingly concerned with instrumental values. It has been a question of the professionals having to prove their case through 'good stories' and better statistics. It is now vital to re-establish the full set of values and to take account of the voice and interests of the public.

RH: This is something that should be of concern to the politicians as well as the professionals. Politicians talk about accountability, but what they need is democratic consent. By the same token, in order for professionals to be able to address the politicians, what they need is the engagement of the public.

JH: Cultural value provides a means to understand what it is that the public values about heritage, what their interests are and how those interests can best be served by professionals and politicians alike. That means paying attention to institutional value in particular. You will hear later from the Work Foundation, who, along with Mark Moore, the theorist of public value, stress the need for public bodies to create what is called an 'authorising environment'. Building institutional value is one of the primary means by which organisations can generate such an authorising environment. But does that imply that heritage organisations should be ruled by public referenda and popular plebiscite? The answer is no. Cultural value gives equal weight to intrinsic value and to the legitimate exercise of professional expertise. There will be occasions when the public interest – and particularly the interests of future generations – will be best served by professionals using the authority of their expertise to contradict the short-term public will.

RH: Through compulsory taxation, and the voluntary self-taxation of the National Lottery, the heritage is funded by the public, but their approval is more important than their pound coins. Unless we reconfigure the relationship between professionals, politicians and the public we will be facing not merely a democratic deficit but a real crisis of legitimacy. That is why today's conference is so important: placing the emphasis not on an internal debate between funders and funded, between professionals and politicians, but putting the public interest at the centre of the discussion.

JH: The way to do this, we argue, is to adopt the language and methodologies of cultural value. This will benefit the politicians, who will gain a stronger mandate for their support for heritage, without detriment to their function as the guardians of the public purse or to their legitimate interest in securing the maximum ancillary benefits in terms of broader policy goals.

RH: It will benefit the professionals, because through recognising and articulating institutional value – and deciding for themselves how their particular institution can best generate it – they will be able to re-validate themselves and the legitimacy of the professional role. Institutions, and those who serve them, will regain the confidence to assert those intrinsic values that the political discourse of the past thirty years has virtually driven underground.

JH: And it is those intrinsic values that draw in the public, upon whom political and professional legitimacy depends. The public will be better served by a system that takes greater account of their needs, that is aligned to their perception of the value of heritage and that gives them a voice. There are three sides to this question: what we are offering is a common answer.

Capturing the public value of heritage: looking beyond the numbers
Accenture

Introduction

In association with the National Trust, Accenture has recently applied its Public Service Value Model (PSVM, patent pending) to offer a solution to the problems faced by the heritage sector in measuring the value of its activities. In advance of more extended publication of this work, this paper gives a summary insight into the workings of the PSV tool, introduces some of the findings from the analysis and explores its applicability to the heritage sector as a whole.

Traditional approaches to performance measurement have various benefits, including the ability to define the relationship between input resources and output results. What they fail to do, however, is take proper account of what constitutes 'value', both in general for the public sector and specifically for the heritage sector. These shortcomings include a:

- concentration on quantified outputs, not on qualitative outcomes
- lack of focus on citizens – actual users, local communities and the wider population
- failure to balance outcomes with cost-effectiveness

There have been a series of efforts in recent years to address these shortcomings through the application of concepts of public value to public services. Mark Moore at Harvard University (Moore 1995), Demos (2004) and the Kelly *et al* (2002) have all made attempts but have not succeeded in bridging the gap between 'value' and performance measurement. The key question for us now, therefore, is to find a model of value that combines the rigour of traditional economically derived methodologies and the sensitivity of the less precise values-based approach in a way that is valuable and meaningful for the heritage sector.

At Accenture we believe that value in heritage has two aspects: 'intrinsic' and 'use'. Intrinsic value is made up of 'soft' benefits inherent in people's experience of heritage and incorporates elements such as aesthetic quality and historical and cultural significance. By its very nature, intrinsic value is difficult to measure. Traditionally, it has been measured by experts but initiatives such as Citizens' Juries are changing views as to how this value is defined. Intrinsic value might best be captured by the amalgamation of a series of judgements made by experts, citizens and local communities. However, one of its potential drawbacks is that it may be difficult to improve intrinsic value such as through operational activities carried out by heritage managers. Intrinsic value has not been explored in detail thus far.

Use value is much easier to measure because it is made up of more tangible benefits. It can be broken down into a series of values or outcomes (such as educational, economic, community), which can in turn be captured through the quantified measurement of indicators such as visitor numbers, user satisfaction and proportions of participants from target social groups. It is possible to improve performance in delivering these values as there is an explicit link between operational activities that are measured and the value they contribute to citizen-focused outcomes. This is an area we have developed further in association with the National Trust; a more detailed paper on this work will be published later in 2006.

The Accenture Public Service Value Model

The Accenture PSV Model combines an assessment of social, economic and environmental outcomes and the way in which they are achieved to identify the public value generated by publicly supported services and assets. It has been adapted from the established commercial principles of 'shareholder value' by shifting the focus to the perspective of the citizen as the investor and key beneficiary or stakeholder. In trying to measure and understand value in the heritage sector, the model has proved to be a helpful tool. Its primary strength is that it can balance the quantification of citizen-focused outcomes with a measure of the cost-effectiveness with which these are delivered. Through discussion with the managers of individual heritage properties it also identifies the operational approaches and practices that appear to have the greatest effect in driving improvements in value.

To investigate the applicability of the PSVM to the heritage sector, the model has been applied to a number of National Trust case-study properties, the results from two of which will be briefly discussed: Montacute House in rural Somerset and Sutton House in east London. The case studies show the benefits of specific management decisions and enable some wider conclusions to be drawn for the overall management of the Trust's asset portfolio.

At the outset, a range of outcomes and sub-outcomes is identified (Fig 9). These capture the high-level economic, social and environmental value and benefits delivered by the heritage sector. A battery of measures, known as metrics, is further identified through which these outcomes can be quantified. These metrics are weighted in terms of their contribution to each outcome. The high-level outcomes are themselves weighted in terms of the contribution to the overall public value delivered by the property. In conjunction with the National Trust, three possible outcomes were identified: optimisation of the user experience (ie visitors); impact on the local community (ie the local population); and impact on the wider population (ie the country as a whole). Putting all this together creates a possible heritage PSVM that can then be applied to the case-study properties, Montacute and Sutton House.

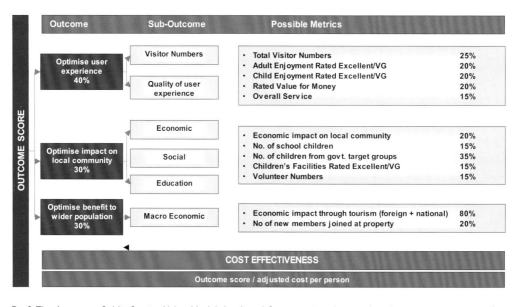

Fig 9 *The Accenture Public Service Value Model developed for measuring the social and economic outcomes of National Trust properties.* © Accenture

Applying the model

Montacute is a magnificent Elizabethan stone-built house with a fine art collection, garden and park. It places an emphasis on increasing visitor numbers and providing a quality education programme. Here, application of the PSVM showed that the overall outcome score for the property had remained broadly constant over the period 2000 to 2004 (Fig 10). At a more detailed level, however, it was apparent that the employment of a new educational officer had significantly increased the local community score. By contrast, a slight fall in visitor numbers and recruitment of new National Trust members in 2003 resulted in a modest reduction in the user score.

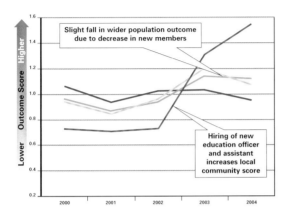

Fig 10 Montacute: Public Service Value outcome scores 2000–2004. © Accenture

Sutton House (see Fig 2) was built in 1535, when Hackney was an unspoilt rural village. Because it now lies at the heart of an area of complex multicultural urban community, the National Trust provides it with support to fulfil its aims of promoting social inclusion and of reducing barriers to access that heritage properties traditionally experience. At Sutton House, the PSV analysis (Fig 11) suggests that improvements in value creation can be linked to a number of developments at the property. In particular, the initiation of free entry days had the effect of increasing visitor numbers. Although this resulted in a corresponding rise in expenditure, the result was an overall increase in the value-creation score for the house.

Fig 11 Sutton House: increased visitor numbers 2000–2004. © Accenture

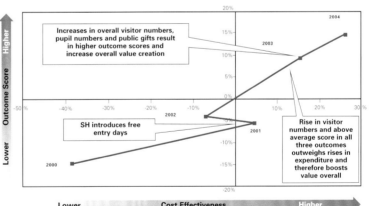

Evaluating the results

The analyses undertaken so far on behalf of the National Trust have demonstrated the potential of the PSVM to assist the heritage sector by:

- defining and focusing on citizen-centric outcomes
- helping sites prioritise their goals through weightings
- offering a decision-making framework to enable better allocation of resources
- offering a way to communicate performance, motivate employees and encourage data collection

- providing a valuable tool for developing strategic plans
- instilling a new value mindset within the organisation.

The case studies have also shown how the model can help to quantify and drive results by:
- linking outcome achievement to cost-effectiveness
- comparing performance over time
- identifying the drivers for achieving high performance
- evaluating spending decisions and investment trade-offs based on projected future performance trends
- managing external relationships.

In summary, we would argue that the PSV methodology has four key benefits:
- it enables a focus on the citizen, both in terms of the organisation and delivery of services
- it encourages a move from target-meeting to value-optimising behaviour
- it promotes understanding of how to maximise value through the identification of value drivers
- it strengthens organisational accountability.

Like all methodologies, the PSVM also has some limitations, which we need to clearly recognise.
- The PSV approach is relative and largely sector-specific. It does not allow us to make definite judgements on whether an agency is good or bad compared with those in other sectors. Similarly, intra-sector comparisons should only be undertaken with caution.
- The analysis is only as good as the data used. A lack of data, or a reliance on inaccurate information, can lead to an incorrect or inadequately comprehensive view of performance.
- Multi-causality means that it can sometimes be difficult to identify and distinguish value drivers with precision.
- On its own, a PSV analysis is not a basis for assessing the intrinsic value of heritage.

Conclusion

The 'value' of this model for the heritage sector is thus threefold. First, by examining the link between outcomes and cost-effectiveness, the heritage sector can demonstrate the value it delivers for the resources it receives. As a result the sector can demonstrate how it contributes to the government's overall social and economic objectives – the common goals that sit above each of the major areas of public expenditure and government activity. By explicitly linking its own outcomes to these overall public policy objectives, the heritage sector should be in a much better position to make its case for its fair share of resources in forthcoming rounds of decisions on public expenditure.

Secondly, by encouraging all its many public and private institutions to maximise use value, the sector has the potential to ensure that it is achieving optimal public participation and current benefit from its assets. And thirdly, by focusing on both use and intrinsic value, the heritage sector might be better placed to manage its wide portfolio range. For example, it could be argued that heritage sites that are in greatest need of public support might have high intrinsic value, but low use value (in spite of management efforts). Equally, it could be argued that those heritage sites with both low intrinsic value and low use value might more sensibly be closed down as managed visitor attractions and instead turned over to other kinds of use.

Heritage, democracy and public value

Ricardo Blaug, Louise Horner and Rohit Lekhi
The Work Foundation, in association with the Research Republic

The discussions in this session of the conference have focused on how to show what it is that is particularly valuable about heritage. Value for money is important, but the public bodies represented here at this meeting are clearly providing other kinds of value that cannot be so easily quantified by New Public Management, with its emphasis on technical efficiency and the public as 'consumers' who need to be 'satisfied'. Public value must therefore somehow articulate the distinctive type of value produced by a public-oriented service – one that reconnects public bodies with the public they are there to serve.

The Work Foundation, in association with the Research Republic, has been researching how the public value concept might be applied in the current context of UK public service reform, and to address problems in existing approaches to public value. In this, our research has covered policing and local government as well as the arts, culture and heritage sectors.

'Public value' is a tricky concept. There are many meanings of 'public', and still more of 'value'. The former can refer to public goods, to social capital, civil society or the public sphere. 'Value' is a term that is equally fraught. To some it means economic value – how much a product or service is worth relative to other things as indicated by its price. Value can also relate to preferences and satisfaction with a particular service at a specific point in time. Finally, values such as security and integrity derive from moral and ethical debate and will always be hotly contested. When it comes to thinking about the value of a historical site, all of these approaches to value have a bearing on decisions about what to conserve or to leave obsolete.

As Kelly *et al* (2002) have argued, public value must, in some important way, reflect what the public values. Living in a democracy, they suggest, this should come as no surprise to us. A public service, they show, has ends and values that are authorised by the public. By authorisation, we mean 'agreed to' or 'owned by'. The purpose of a public service is thus to create public value, and to do so efficiently.

Heritage, of course, requires a careful balance between stewardship, or trusteeship, and an orientation to what the public wants. Services must respond and react to what the public wants, and to do so across the design, provision and evaluation of services. This is no more than to repeat the importance of localism, and the move away from centralised standardisation. This is not to say that standardisation does not have a role to play, for example in driving up minimum standards. Rather than continue with the age-old tug-o'-war between the forces of centralism and the forces of localism, the shift we are advocating here is about getting organisations to reconnect with their public – the public might be a global one, as with most World Heritage Sites, or it might be those living within walking distance of a royal park. Either way, understanding the public is the key, rather than simply debating who should have the right to make decisions on the public's behalf.

In order to find out what the public values, service providers must interact and engage with the public. This engagement is a two-way exchange of knowledge, for providers and the public here educate each other. Not only do service providers seek to shape public preferences, for example providing expertise and knowledge about the history of the object or building that the public would otherwise not have been aware of, but they also learn from the public, from its opinions, interests, expertise and collective clout. In the language of public value, providers must interact with their 'authorisation environment' to find out what the public values. It is the public that must 'authorise' the value to be pursued.

This element features strongly in the Work Foundation's conception of public value. The public authorises what is seen as valuable. Again, it is worth remembering that the 'public' here really means many overlapping 'publics', a teeming mosaic of interests variously pursued by associations, institutions and the media – indeed, the complete 'authorisation environment'. Yet it is the public authorisation of goals and evaluative criteria that distinguishes public value from economic value, not simply the absence of a price mechanism.

To orient to what is valued by the public is hardly a new concern. At least since the 17th century, we have aspired to be democratic, to represent and respond to public preferences, even if we have struggled to do it well. Nor do the problems with democracy arise here for the first time. To respond to what the public wants is not, for example, merely to pander to the lowest common denominator or mob rule. When the 'public' mistakes a paediatrician for a paedophile, as occurred following the *News of the World*'s campaign against the latter in 2002, we forcefully argue that there is, actually, a difference. In democratic theory, a distinction is drawn between a knee-jerk, ill-informed and unconsidered preference and a 'refined' preference (Fishkin 1992, following a first use of the phrase by James Madison in 1788 in 'The Federalist No. Ten', quoted in Hamilton *et al* 1961).

It is worth noting that there is nothing especially 'refined' about refined preferences. It is not a way of reasserting the refined, high-end tastes of the middle or upper classes. But it does require some thought to have taken place among the public. It means that something took place beyond a mere reactive jolt, that some consideration or discussion went into a decision that was made in the public's name and with its money. A preference is more influential if it is well informed, educated, negotiated, discussed, chewed-over and then given. The public might still hold the same views, but the fact that these were arrived at via a defensible process significantly increases the public's demand to be heard. To an economist, this is about addressing an asymmetry of information. But it is also more than this, for it gives a democratic justification for why the public needs to better understand what the public purse provides on its behalf.

A notion of refined preference can, of course, provide an 'easy out' for elite decision-makers seeking to exclude the public from involvement, for they can claim – as did Robespierre and Lenin – that their special knowledge makes them the keepers of the public good. Only when professionals seek to refine public preferences and are willing to have their own preferences refined, is the required balance achieved. In this way, heritage is, as Fiona Reynolds, Director-General of the National Trust, put it during a discussion session at this conference, collectively 'discovered'.

A service should be responsive to the public, yet also seek to refine public preferences. This implies:

public value = responsiveness to refined preferences.

Such a definition lends clarity to the Secretary of State's call for an articulation of the value of heritage and the importance of the policy 'conversation'. It acknowledges the role of specialised knowledge (stewardship) and the interests of the public (again, this process of refinement also affects professionals!). Furthermore, refinement of preferences is a way of overcoming the often ill-conceived and even maligned notions of the public. It requires decision-makers to seek greater understanding of the public in whose interests they are supposed to act, and to furnish it with information, access and leadership to that end. In this way, decision-makers win legitimacy, trust and support.

We can then proceed to use public value to:

- show the value of a service – and how it reflects the needs and values of the diverse publics it serves
- clarify the core mission and cascade that mission throughout an organisation (using public value as a management tool)
- educate the public about what an organisation does
- get publicly authorised goals, evaluative standards and support for activities and initiatives. It is this that confers legitimacy, both to receive public funds, and to act as a steward of 'public' interest
- motivate organisations to orient to the public, to interact with their authorisation environments, to learn about them, to educate and respond to them
- argue for, and justify, the allocation of resources – using the rhetorical power of public value
- demonstrate, and measure, the creation of value in public service.

In seeking to quantify public value, we enter the world of performance indicators, the centralised specification of evaluative criteria and, too often, the loss of connection with the local public. There is little question that measurement can both create and destroy public value. Here, however, we must state the *kind of value* created by the heritage sector. This entails a quantification of the quality of interaction with the authorisation environment To measure public value is to measure *responsiveness* to, and the *refinement* of, preferences.

This cannot be an absolute value, any more than the meaning of life can be finally found to be 42. Public value is not a standard unit, applicable everywhere. It cannot compare, for example, outputs across different contexts. This is because public value, in being responsive, is different in different locations. Nor can it assign a public value unit of value in place of economic value. It cannot, therefore, 'evaluate' a particular item of heritage. Rather, a Public Value Performance Indicator would measure an organisation's capacity to create *public value*. In other words, it would measure the capacity of an organisation to refine and respond to public preferences. This would make meaningful comparison possible between organisations of different size and located in different contexts. A small local museum may not be as big as a historic royal palace, but the former could have a greater capacity for, and much better methods of, creating public value.

In seeking to quantify public value and use it effectively, we cannot anticipate the outcome of the democratic interaction. We cannot impose values without interaction with the authorisation environment – the public, the media and politicians. We cannot jump to the end of the process and simply dictate what we consider to be the relevant values and goals. This is a democracy, and though somewhat noisy and impractical as a method of decision-making, it remains the best way to gain authorisation, legitimacy and the noisy excitement of innovation.

Elements of the Public Value Performance Indicator we propose would therefore include the quantification of responsiveness and of refinement. In each case, the number and quality of interactions would be recorded. Thus:

Quantifying institutional responsiveness	Quantifying refined preferences
■ Deliberative engagements	■ Educational initiatives
■ Consultations	■ Information dissemination
■ User participation	■ Transparency
■ Consumer feedback, surveys, etc	■ User participation
■ Policy adaptation	■ Ongoing evaluation
■ Leadership (from below – engaging the public)	■ Leadership (from the front – shaping preferences)

Measuring and valuing in this way will entail an illumination of what people in the heritage sector already do. Yet it would serve, among other things, to highlight the importance of educational initiatives, to favour certain models of leadership and to distinguish between consultation and deliberation.

In order to examine and rehearse the notion that public value is institutional responsiveness to refined preferences, it is worth considering two examples. In the first, we witness the recent, sudden and celebrity-driven interest in school dinners (Fig 12). This example shows how public preferences can rapidly shift, and how public value can be created by responding to that shift.

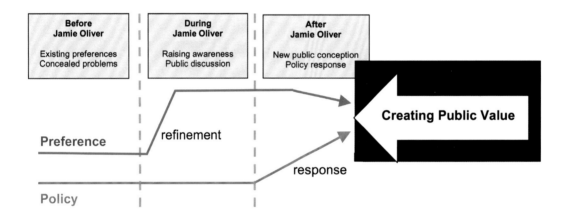

Fig 12 Responding to public preference. Example 1: the sudden celebrity-driven interest in school dinners leads to a change of government policy. © The Work Foundation

The second example concerns the way in which the National Health Service persisted with MMR vaccination despite public panic (Fig 13). In this case, and in contrast to the school-dinners situation, there was a studious lack of responsiveness on the part of government. Instead we saw professionals not pandering to ill-informed public preferences; seeking to refine and educate public preferences; allowing time for public debate before reacting; and defending actions with good reasons (here, better science).

This has clear similarities with heritage, where professionals are partly stewards. Professionals thus need to justify (legitimate) their right to say what is really valuable, and the right to question what the public says it wants.

Public value thus shows the distinctive value of heritage: that it creates public value. This clarification should enable organisations to argue more effectively for resources, on the grounds of their high capacity to create public value through valuing what the public thinks, wants and needs. Such an approach helps balance stewardship with a genuine attention to public interests, and it counsels responding to, yet also informing, the public.

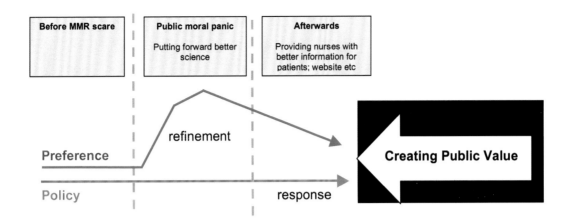

Fig 13 Responding to public preference. Example 2: the National Health Service persists with MMR vaccination despite initial public panic. © The Work Foundation

It articulates the relation between excellence and public involvement called for by Tessa Jowell, and makes this balance real at the organisational level. Public value is more than a measurement system for it also affords strategic guidance about what to do.

Adopting this approach, however, does not mean that tried and tested tools of performance measurement and economic value, including for example concepts such as 'willingness to pay', should be thrown away. Rather, the public value approach argues that public managers seeking to understand the value they create should ensure that these approaches take account of what the public thinks and ways in which those opinions need to be better informed.

What is distinctive about heritage, therefore, is its capacity to create public value. In turn, what is distinctive about public value is that it enables heritage to claim democratic legitimacy.

Discussion 1. Public value: how useful is it as an idea?

Chair: Nick Higham
(Arts and Media Correspondent, BBC)

Panel

Ricardo Blaug (Work Foundation)

Robert Hewison (Demos)

John Holden (Demos)

Fiona Reynolds (Director-General, National Trust)

Mark Friend (Controller of Broadcast Strategy, BBC)

Greg Wilkinson (Accenture)

The issues

- Is 'public value' a helpful concept in a world dominated until now by more rigorous systems of economic cost-benefit analysis?
- Can it be usefully applied to the heritage and how relevant is it likely to be to heritage organisations in their dealings with government?
- What implications does it have for the way in which heritage organisations behave?

The words below are not necessarily exactly those that were spoken, and are intended instead to capture the overall flavour of the discussion.

Nick Higham (Chair)

Public value can be defined as what citizens get from the activities of public bodies. There is a certain amount of popular cynicism that it is something that can actually be measured. However, it is also important to recognise that the concept of public value arises from a genuine desire by government to understand the things that matter to people. Perhaps even more relevant is the practical reality that working with public value is going to be vital for any organisation that wants to win funds from the Treasury or the Lottery. At the moment the government thinks that many heritage bodies are still not trying hard enough to understand public value, are not yet talking properly to the public, and have still not learnt how to sell themselves effectively to the Treasury. And we should all remember that John Prescott would rather listen to *people* than *experts*.

Ricardo Blaug

Intrinsic value is not something that it is easy to *measure*, it can only be taken on trust.

Robert Hewison

Intrinsic value is something that has to be *articulated*, not *measured*. The vocation of the heritage professions is therefore to articulate the public voice about what matters. More immediately, their moral task is to transmit that message to the government's 2007 Spending Review.

Fiona Reynolds

In the National Trust we have already come a long way and people *are* now engaged in expressing the value of historic places in new ways. Our job is to communicate those views to the Treasury and the Heritage Lottery Fund in terms of bids for resources.

Greg Wilkinson

If the sector wants to influence Treasury, it has to understand that there is no alternative to hard evidence – but it also has to realise that it will be facing a difficult task, as most of its own evidence is qualitative, not quantitative.

Robert Hewison

There is an argument that the National Trust has reached a point at which it is taking too much money from the Heritage Lottery Fund for big national projects like Tyntesfield.

Fiona Reynolds

Education is one of the most powerful means of capturing public value, which is why education is the key justification for Tyntesfield. People are not just being *told* things, they are actually *involved* in the project. And it is important to realise that Tyntesfield is not just about the past, but the future too.

Mark Friend

The BBC has been using the concept of public value to legitimise its bid for the renewal of its charter. It has been a sequential process of proposing ideas, listening to the response of the public, refining the ideas on the basis of that listening, and then listening once again to reactions to our refined ideas. In the old days the BBC measured its success in quantitative terms of audience *share*; today the focus has shifted to a more qualitative assessment of audience *reach*. The challenge is to measure the enduring impact that the BBC is having on people and their lives. Establishing a relationship with the audience is the new priority.

Delegate

How do we respond to specialist interests in this new environment of popular value; how does the process of refining public value avoid crushing minority opinion?

Fiona Reynolds

The National Trust has never tried to please everyone all the time. Some places are consciously managed to respect specialist interests; the Trust would be very worried if minorities were excluded.

John Holden

Sustaining the breadth of heritage is as vital as sustaining biodiversity.

Ricardo Blaug

It is false and naïve to assume that the process of refining public value will eventually produce a single uniform consensus about what matters. In a democracy, the majority voice will always tend to predominate. Disagreements will never go away and it is important that public policies should properly reflect and respect them.

Delegate

It is important to distinguish two distinct purposes of public value work. On the one hand it can be used as a tool for winning money via advocacy to the Treasury; on the other hand it can be used to help individual institutions perform better. The second of these is laudable, but the first is worrying because it implies a government that lacks confidence in its own judgement and needs the endless reassurance of 'evidence' before it can make up its mind.

Greg Wilkinson

It would be a mistake to ignore the advocacy opportunity offered by public value – heritage is actually in quite a good position to justify itself.

Delegate

Heritage places are often those where people want to get away from things and be on their own – an importance that is totally at odds with value measured in terms of crude visitor numbers. It is also important to remember that appreciation of the value of a particular class of heritage asset usually begins with a tiny group of enthusiasts, and only later becomes mainstream. In our concern to capture the opinion of the majority we must not lose sight of the insights of the pioneers.

Delegate

Public authorisation at the centre seems limited when compared with what happens on the front line. We should also remember that the real reason that government responded to Jamie Oliver's school dinners campaign was nothing more than fear of public opinion.

Mark Friend

The role of the media is vital in transmitting values, as was evidenced by the popular responses to programmes like *Restoration* and *Springwatch*. It is therefore very important to develop a constructive partnership between heritage professionals and media.

Fiona Reynolds

Heritage is not static; it moves with society. The media can document that change as it happens – and it's not just people who are changing; the professionals are too!

Delegate

Local government used to be driven by public value in terms of its conservation priorities, and it will be again after the implementation of the government's Heritage Protection Review – but will it have the capacity and resources to listen and respond to popular opinion in the way that is being advocated at this conference?

Fig 14 At the end of the first day of the conference, Urban Roots Dance Project performed songs from Porgy and Bess and dance routines celebrating their own street culture. The performance grew out of a project funded by the Heritage Lottery Fund. Members of St Mary's Youth Club in Islington wanted to learn more about the origins of UK street dance and to use this as a tool to explore their own cultural heritage. After making a successful application for £25,000 from the Heritage Lottery Fund's Young Roots programme, the club ran a ten-week summer programme tracing the origins of UK street dance right back to African and Caribbean dance, traditions and culture. The project finished with a sold-out performance at the Hackney Empire. © Heritage Lottery Fund

2 The Instrumental Benefits of Heritage

Baroness Andrews
Sustainable Development Minister, Office of the Deputy Prime Minister
Sustainable communities: the economic, social and environmental benefits of heritage

David Throsby
Professor of Economics, Macquarie University, Sydney, Australia
The value of cultural heritage: what can economics tell us?

Sue Wilkinson
Director of Learning, Access, Renaissance and Regions, Museums, Libraries and Archives Council
Capturing the impact of museums on learning

Julia Thrift
Director, CABE Space
Public space: public value

Heather Garnett
County Councillor and Heritage Champion, North Yorkshire County Council
Heritage on the front line: the role of a heritage champion in North Yorkshire

Kate Clark (chair)
Deputy Director, Policy and Research, Heritage Lottery Fund
From significance to sustainability

Discussion 2
The instrumental benefits of heritage: how are they measured?

Sustainable communities: the economic, social and environmental benefits of heritage

Baroness Andrews
Sustainable Development Minister, Office of the Deputy Prime Minister

I understand that the young people from Castleford are giving you a practical demonstration of how heritage is more than just buildings – it is people, their history and their culture (see Fig 33). I know that Yvette Cooper, my fellow Minister at the Office of the Deputy Prime Minister (now Department for Communities and Local Government), would be really proud of the performance by some of her younger constituents. She has supported the Castleford Project from the start and she was pleased to hear that the project is now confident enough to take to the national stage as a demonstration of how important heritage is to creating a sense of community – a sense of place.

When people ask me to describe what the Office of the Deputy Prime Minister does, I proudly say it is the Department of Place – the department whose task it is to make places work for people, to create sustainable communities. I am therefore very pleased that you selected this venue, the Royal Geographical Society – or as I might put it if I worked here, the Royal Society for People and Places. There are many reasons why this is an appropriate venue. For me, it is profoundly evocative of a turning-point in the history of knowledge and the learned societies that underpins the values and the promise of this conference itself. The early 1830s saw science transformed from a private and fashionable pursuit into a popular and democratic movement – not only the opening-up of the Royal Society, but the formation of the British Association for the Advancement of Science, the Statistical Society and the Royal Geographical Society, and, indeed, something which is certainly part of our cultural heritage, the Royal Horticultural Society. But in terms of the Royal Geographical Society itself, then, as now, men (although now there a few more women) of science were concerned, as we are, with understanding the state of the planet, the undiscovered possibilities and fugitive relationships between science and ethics, and the link between geological past and a sustainable future.

So – this is an appropriate venue for a conference that is exploring the public value of heritage, by which we mean not only the built and natural environment but also the collective and cultural memory, and how that in turn can inform and guide us through a very challenging climate of change. For that and for many other reasons, I was delighted to accept this invitation to share a platform with colleagues across government, and for the opportunity to reflect on the unique role and opportunity we in the Office of the Deputy Prime Minister have to design a future in which the past is not discarded or diminished but takes its rightful and creative place at the heart of future communities.

Fig 15 The Whitefield area of Nelson in Lancashire, an intact 19th-century industrial townscape in which terraced housing is intimately associated both with the former textile mill and the social amenities of church, school and shops. In response to opposition from the local community and English Heritage to the original plans, the area is now being refurbished by a partnership led by the Elevate East Lancashire housing market renewal pathfinder and including the North West Regional Development Agency, English Partnerships, English Heritage and representatives of the local community. © English Heritage

To do that successfully, there are three things in particular that we need to do:

- to break down the perceived and real barriers between 'heritage' and development
- to show how in practice we are putting a public value on heritage in the creation of sustainable communities, both in terms of private space as well as in the public realm
- to show how the public value of heritage is helping people to take control of change through community regeneration and renewal.

This means taking on some powerful mythologies that mirror our three tasks:

- first, it is a myth that all development is intrinsically inimical to an appreciation of heritage
- secondly, it is a myth that regeneration must be at the expense of the community character and continuity
- thirdly, it is a myth of grand proportions that in our passion for house-building we will not only build over our past but sacrifice the future too, as we lay waste to the countryside and the natural environment.

These mythologies, taken together, suggest that we believe communities that are not connected to their own history and heritage can somehow thrive. Of course they cannot.

Sustainable communities

The strapline for my department is: creating sustainable communities – ie living communities that are sustainable because people are proud to live there; communities where people feel they belong because they identify with the place and their neighbours; communities where people want to bring up their children.

Let me start with the tools that we have to hand. Of course, one of the most powerful and creative tools we have to create sustainable communities that are a delight to live in is planning itself. Indeed, the Planning Act of 2004 creates a statutory purpose for planning – that of contributing to sustainable development. Sustainable development is at the heart of sustainable communities, and there is no way we can achieve this unless we recognise and expand on the role our historic environment has to play. A planning system that works well is one that ensures that communities have what they need to thrive. But it is also one which is intent on listening to the community around it and to what people say about what makes that community the place they want to live in. Very often, what people grasp as most important to them is the elusive feel of an area, its continuity, the way the past and the present reflect each other.

We have put that squarely into planning policy. *Planning Policy Guidance 15 (PPG 15)* states that 'the physical survivals of our past are to be valued and protected for their own sake, as a central part of our cultural heritage and our sense of national identity' (DoE/DNH 1994). In terms of new build, *PPG15* states: 'the design of new buildings intended to stand alongside historic buildings needs very careful consideration. In general it is better that old buildings are not set apart, but are woven into the fabric of the living and working community.'

The 2004 Act introduced reforms that give local planning authorities further tools to ensure that these policies become reality. The new Local Development Frameworks comprise a folder of documents for delivering the spatial strategy for the area. These should set out clear policies for the preservation and enhancement of the historic environment. The new Statements of Community Involvement will enable the local community to have a stronger voice in the planning system – local authorities must, at the outset, make clear what process and types of involvement will be gone through to involve the community. The statement should be a clear public statement enabling the community to know how and when they will be involved in the preparation of local development documents and how they will be consulted on planning applications.

The point is that only if we can help people to feel part of – indeed involved in maintaining – their community's heritage will they then truly value it. That does not mean over-protection or denying change at all, but ensuring that our historic buildings (whether they are domestic or public), our parks, our historic green spaces, our industrial history and our natural habitats have their continuing place in the heart of the community. We must do this not just in terms of providing more housing and creating new communities but also in regenerating traditional communities. So I want to invite this audience to join me on a journey on how things are changing and how in the widest sense, the public value of heritage is being expressed across the country, in the growth and renewal of communities.

Barker and heritage

For decades, under successive governments, this country has built too few homes. Over the last 30 years, as demand for new homes has increased by one-third, housing construction rates have halved. At the end of last year the Chancellor and the Deputy Prime Minister responded to Kate Barker's review of housing supply by announcing our commitment to increasing the rate of house-building from 150,000 per year today to 200,000 by 2016. This will help to bridge the gap between demand and supply for homes.

Building these homes will be a challenge, not only because we must meet the highest environmental challenges – and new build gives us the best chance of doing that in many ways – but also because we are serious about building quality communities and about using land wisely and well. That means getting across to local authorities, developers and housing providers that we are *serious* about the quality of the public realm, the design of townscape and public spaces, and the design of houses.

Already 70 per cent of new homes are built on previously developed land – building on green fields is only ever a last resort. We have also increased the number of new homes that are built per hectare from 25 to 40 since 1997. By doing this we can build 1.1 million homes on less land than the previous government set aside for 900,000 homes and save 5000 hectares, an area of greenfield land larger than Norwich. This is protecting our natural environment – the landscape that we all treasure.

Part of improving the quality of these developments must be giving the community access to its heritage – it is about making sure we see economic regeneration as part of a wider picture. It is also about recognising that local distinctiveness can add value. The *Streets for All* regional series of guides produced by English Heritage and supported by government encourage those responsible for streets to reflect local heritage and to design public spaces to enhance their setting. South-East England and the Thames Gateway have a rich and varied tapestry of materials, traditions and characteristics to draw on in creating a distinctive and appealing character for new development.

New communities are growing out of different sorts of heritage, too. Look at the Thames Gateway. This area, which has been identified for growth, has a rich and complex heritage that left us many brownfield sites, but David Miliband has talked about it in terms of offering a greenfield opportunity for innovation. This is an unparalleled opportunity because, as English Heritage's recent Land Characterisation Study of the area has revealed, it is not an environmental vacuum but a varied and fascinating historic environment, part of the biography of London, and as such to be celebrated in terms not only of its maritime heritage but also of its historic rural landscape.

When the dockyards closed, the area with the most historic interest became the responsibility of the Historic Dockyard Trust while Chatham Maritime have taken responsibility for 140 hectares of the site. They have worked together and created a real sense of community based on the heritage of the site – the two landowners have involved the

Fig 16 Chatham's 180m-long drill hall has recently been converted into an educational resource centre.
© Jim Higham, University of Kent

residential and commercial community. The UK's highest concentration of listed buildings has been reused and the fabulous 180m-long drill hall is an educational resource centre – the historic dockyard teaches 13,000 schoolchildren a year about the heritage of the site (Fig 16). In total the redevelopment of this site has opened up the history of the area to 2 million people. It was a worthy finalist in last year's Deputy Prime Minister's award for sustainable communities.

Parks

Let me take another example of our heritage that transforms lives and expectations. As any good architect will tell you, what gives a place character is not the house you live in but the place: the street, the neighbourhood, the local park.

We have had to invent a clumsy hybrid word – liveability – because there was no word to sum up how people interact with where they live and how that influences their whole attitude to their surroundings. Parks are often central to forming that attitude. The best have always been places of great liberation and community enterprise – for play, for picnics, for recovery, for recreation and for meeting and making friends.

We are building new parks and spaces that link the past and future. Look at the centre of Sheffield. The regeneration of the Peace Gardens has given the historic civic buildings – those reminders of Sheffield's Chartist history that include the great municipal riot of a town hall – a new context that enables them to serve their original function as the focus of a proud city. Today Sheffield is a more diverse community and the regeneration reflects that.

David Lammy has talked at this conference about the need to broaden our definition of heritage and we need public places to do that. Parks are changing as communities change. Lister Park in Bradford – a recipient of funding from the Heritage Lottery Fund and a Green Flag winner – is an excellent example of how a historic park has continued to grow with its community. The original park that was created in the late 1890s now includes a new Mughal Garden, a successful mixture of Islamic and Hindu styles reflecting the diverse communities in that area of Bradford and a good example of the adaptability of spaces (Fig 17). This change does not just apply to parks but to any public, shared space. I visited Southampton recently and saw some fantastic sculptures that celebrated the historic maritime past of the area. These sculptures not only looked great but they

Fig 17 A £3.2 million grant from the Heritage Lottery Fund was used to transform Bradford's decayed Lister Park into a popular destination for the city's culturally diverse population. © Bradford Council

had really pulled the community together. People in the area had something of the past to identify with and the community had played its part in designing them.

I am pleased that my department is working with English Heritage and the Heritage Lottery Fund to promote the value of, and the best practice in, the care and upkeep of historic parks. I am particularly pleased with the Heritage Lottery Fund's significant investment in parks. It has already invested £400 million in the regeneration of public parks, and last week announced a partnership with the Big Lottery Fund for a further investment worth approximately £160 million. What is more, there will be an expectation that these parks are brought up to the standard of the national Green Flag Awards, which we support.

There is still a great deal to be done to ensure that parks are at the centre of community life. Our support for the Green Flag Awards is in part an attempt to highlight the icons of good practice, the places to emulate, the places other communities want their parks and open spaces to be like.

Community involvement

The way in which parks mobilise people takes me on to my third point – sustainable communities are only sustainable if they command the loyalty and passion of the people who live there. All the evidence is that from Lewes (where I live) to Lerwick there is no more effective way of engaging people in new or renewed communities and neighbourhoods than by engaging with their own history – whether that is cultural or physical.

I'm pleased that there are fantastic examples of the arts being used in the process of building sustainable communities, too. The excellent Creative Neighbourhoods programme supported by the Arts Council and Housing Corporation (Carpenter 2004) shows how the arts can effectively transform people's lives and communities when embedded in housing and regeneration programmes. There is a particular role for young people here. 'Images of Newbold' in Rochdale involved young people and community workers using positive and negative images of their area to improve understanding and cohesion between culturally diverse residents. Not only has the community learnt new skills around photography and digital manipulation, but it also has a better understanding of its place and the different people within it, which has resulted in more active involvement in shaping its future. I'm particularly pleased it reached young people and engaged them – as the local community worker said: 'the children have soaked it up like a sponge'.

Another example is Ladywood, Birmingham, where children have been engaged in capturing the memories of local people in a community in which there has been rapid change and renewal and where streets have been restored. And I have already played tribute to the young people of Castleford – but we must not forget that Castleford has been through some tough times and we should not ignore the economic value of heritage.

Economics

Economic value, of course, is another self-evident argument. We have many outstanding examples of how the very best developments have incorporated the historic heritage of the local site. For example, the Berkeley Homes development of 905 homes on the Royal Arsenal site involved the refurbishment of a number of historic buildings, which helped to raise the quality of the housing stock and provide the area with a unique and attractive setting. The link with regeneration is often economic. Restoring the historic environment creates jobs and supports local economies, as we have seen in the heritage-led regeneration of the Jewellery Quarter in Birmingham, where it has ensured that this area remains a thriving centre for manufacture. The skills that are needed to maintain our heritage – from stonemasons to community tour guides – will also be fostered by putting heritage at the centre of sustainable communities.

Market regeneration and renewal

Economic and social regeneration lies at the heart of what we are doing in areas of the country where communities have been virtually abandoned to their fate – the last part of the journey. Here, the best approaches to economic regeneration are built firmly on the community's identity – slum clearances and the destruction of whole communities should be things of the past.

Let me assure you that despite the lurid headlines, we are not hell-bent on demolishing the North. Instead we are seeking to create a sustainable future for communities that were in crisis. A collapsing housing market has resulted in homes being abandoned and left to become derelict. To be frank, the end of heavy industry meant there was no longer a need for a large local workforce to live in and around the mills and factories of our northern cities, and if people cannot find work they will move on.

The Pathfinder programmes are prioritising the refurbishment of homes (see Fig 15). To date, more than 13,000 houses have been refurbished compared to 4,000 demolished. Demolition is never the first or the only option but sometimes refurbishment cannot provide the larger houses, the gardens, the better aspect, the space and light that growing families want. Indeed even houses that had been renovated in successive and expensive local and national initiatives were lying empty as no one wanted to live in them. The community was struggling to survive – but people are proud of their heritage – and it is imperative that this sense of community is the basis for the regeneration of the community. All the Pathfinders have been encouraged to look at the character of their areas, to assess the 'look' of the buildings and to address the local community's wishes for the future of its historic buildings. You see we have learnt lessons from the past.

I am delighted that English Heritage and the Commission for Architecture and the Built Environment (CABE) are working so closely with the Pathfinders to ensure that we achieve the balance between retaining the character and spirit of place and bringing new life to these communities. Indeed, in collaboration with CABE, English Heritage, the Environment Agency, the Commission for Integrated Transport, and the Sustainable Development Commission have produced a document entitled *Building Sustainable Communities: Actions for Housing Market Renewal* (CABE 2003). Chapter 2 of this publication focuses on heritage and outlines how Market Renewal Pathfinders can positively address heritage as an asset. English Heritage has started to work with both the Merseyside and Birmingham and Sandwell Pathfinders.

Conclusion

I want to finish where I started – with what this building represents. The collective memory. Finding the collective memory can both create social cohesion in new communities that have to establish their identity and also nurture social cohesion even in communities that are facing dramatic change. I have seen how identity around loved landmarks, buildings, parks and cultural activities can keep a community together and enable it to manage change confidently. The power of heritage in building communities is the power of engaging people's imaginations and passions. Without that, communities cannot thrive. And, of course, shared history can bring not just communities but generations together.

To conclude, here is a quote from Chapter 1 of Peter Ackroyd's *London: The Biography*:

> If you were to touch the plinth upon which the equestrian statue of King Charles I is placed, at Charing Cross, your fingers might rest upon | the projecting fossils of sea lilies, starfish or sea urchins (Fig 18). … In the beginning was the sea. There was once a music-hall song entitled 'Why Can't We Have the Sea in London?', but the question is redundant; the site of the capital, fifty million years before, was covered by great waters.
>
> The waters have not wholly departed, even yet, and there is evidence of their life in the weathered stones of London. The Portland stone of the Customs House and St Pancras Old Church has a diagonal bedding which reflects the currents of the oceans; there are ancient oyster shells within the texture of Mansion House and the British Museum. Seaweed can still be seen in the greyish marble of Waterloo Station, and the force of hurricanes may be detected in the 'chatter-marked' stone of pedestrian subways. In the fabric of Waterloo Bridge, the bed of the Upper Jurassic Sea can also be observed. The tides and storms are still all around us, therefore, and as Shelley wrote of London, 'that great sea … still howls on for more'.
>
> *(Published by Chatto & Windus. Reprinted by permission of The Random House Group Ltd.)*

Fig 18 Our shared history – the statue of Charles I was erected at Charing Cross in 1674, on the spot where several of the regicides were executed. It faces down Whitehall towards the scene of the king's death. © English Heritage. NMR

The value of cultural heritage: what can economics tell us?

David Throsby
Professor of Economics, Macquarie University, Sydney, Australia

Introduction

Whether we like it or not, the contemporary world is increasingly being shaped by market forces, as the effects of globalisation continue to spread through the economy and as governments come to rely more and more on policies based on the precepts of neo-liberal economics. In these circumstances, how is the case for some government involvement in the conservation of cultural heritage to be put? It is well known that many of the benefits of heritage accrue not just to individuals but to the community at large and as such are not reflected in market transactions. Is it possible in a market-dominated world to make a case for heritage protection? And can this case be put using language that economic policy-makers can comprehend? In this paper I suggest some basic principles that can guide our thinking about the economics of heritage and lead towards policy formulation in this field that is sensitive to both the economic and the cultural values at stake.

Cultural capital

The term 'cultural capital' is well known to sociologists following Pierre Bourdieu, but in economics it is acquiring a somewhat different interpretation. Economists look upon capital both as a store of value and as a long-lasting asset that produces a stream of services over time. An item of cultural heritage can be thought of as just such an asset. Consider the case of a historic building. It is appropriate to regard any building, historic or otherwise, as a capital asset that gives rise to a flow of services, and that will deteriorate (and hence depreciate) if it is not maintained. But if the building is a heritage building, it can be suggested that it embodies not just *economic* value (which could be realised by putting the building up for sale) but also *cultural* value, some intrinsic or assigned quality which stands apart from the building's financial worth and which reflects some evaluation of its cultural significance. It is this cultural value attributable to such an asset that allows it to be classified as an item of cultural capital, as distinct from 'ordinary' physical capital whose value can be fully captured in economic terms.

To put it more formally, we can define an item of cultural capital as being an asset which embodies or yields cultural value in addition to whatever economic value it embodies or yields. The phrase 'embodies or yields' is used here to emphasise the distinction between the capital *stock* and the *flow* of capital services to which that stock gives rise, a distinction which is fundamental to analysis of any sort of capital in economics. In the case of a heritage building, the asset *embodies* value as a piece of capital *stock*, where that value is expressible in both economic and cultural terms. In turn, the building *yields* a continuing *flow* of services over time, such as the accommodation it provides for tenants or the benefits accruing to tourists who may visit it as a cultural site; these flows also generate both economic and cultural value, which can, in principle at least, be identified and measured.

Why is the concept of cultural capital helpful in the process of formulating heritage policy? I can suggest four reasons. First, the phenomenon of 'capital' is, as noted above, an important one in economics; defining heritage as capital enables the related concepts of depreciation, investment, rate of return, etc to be applied to the evaluation and management of heritage. In so doing one can open up a dialogue between heritage professionals whose job it is to care for cultural assets and economists who are concerned with the formulation of economic and cultural policy. Second, the idea of cultural capital depends on articulating specific forms of value. In particular it draws attention to cultural value as something distinct from (though not altogether unrelated to) economic value. I return to the question of cultural value in more detail below. Third, since capital assets are long-lasting, the notion of cultural capital leads naturally to thinking about sustainability. We are now accustomed to speaking of environmentally or ecologically sustainable development as being a growth path for the economy that preserves the natural resources of the planet for future generations; in exactly the same way it is possible to speak of culturally sustainable development, meaning ways of safeguarding our cultural heritage for the benefit of our children and our children's children. Neglect of cultural capital by allowing heritage to deteriorate, by failing to sustain the cultural values that provide people with a sense of identity, and by not undertaking the investment needed to maintain and increase the stock of both tangible and intangible cultural capital, will place cultural systems in jeopardy and may cause them to break down, with consequent loss of welfare and economic output.

Fourth, it is usual to apply economic appraisal methods such as cost-benefit analysis to public investment in capital assets. Defining heritage as cultural capital opens up possibilities for looking at heritage projects in similar cost-benefit terms. For example, an intervention involving expenditure of public or private funds can be seen as a capital investment project. In such a case, if the asset is a historic building or location and the 'project' is the restoration, reuse or redevelopment of the site, we can suggest that treating the cultural resource as an item or items of cultural capital enables the familiar tools of financial investment appraisal to be applied. But there is an important difference from 'ordinary' cost-benefit analysis: it is (or should be) the time-stream of *both* economic *and* cultural value that is being evaluated and assessed. In other words, the identification of cultural value alongside the economic value generated by the project means that the economic evaluation can be augmented by a cultural appraisal carried out along the same lines, ie as an exercise comparing the discounted present value of the time-streams of net benefits with the initial capital costs.

Economic benefits of heritage

Over the last two decades or so, an increasing amount of attention has been paid by economists to measuring the tangible and intangible benefits yielded by natural environments. Recently these methods have been applied to assessing the benefits of cultural heritage. Essentially these methods involve distinguishing between the direct use benefits of heritage that accrue to those using the assets, such as tourists, and the indirect or non-use benefits that accrue to the community at large. The former can be measured by market transactions, but the latter arise outside the market, and have to be measured by special-purpose studies designed to gauge people's willingness to pay to preserve the heritage in question. These non-use values may relate to the asset's *existence* value (people value the existence of the heritage item even though they may not consume its services directly themselves); its *option* value (people wish to preserve the option that they or others might consume the asset's services at some future time); and its *bequest* value (people may wish to bequeath the asset to future generations). These non-use values are not observable in market transactions, since no market exists on which the rights to them can be exchanged.

Willingness-to-pay studies of various sorts can be conducted at the micro-level to assess the community benefits from conservation of a specific heritage item – a monument, an old market square, an ancient church, an archaeological site, etc. Alternatively they can be applied at the macro-level to find out how much the population at large would be willing to see spent out of its taxes on heritage protection in general. An example of the latter is a recent choice-modelling study carried out in Australia as input into a government enquiry into the built heritage; the study was able to demonstrate that the average citizen thought that not enough was being spent on heritage conservation at the present time. The study provided solid evidence that an increase in Australian government funding for heritage over present levels would be in line with consumer preferences.

So, to summarise, economic evaluation methods, if systematically and rigorously applied, are essential tools in understanding the value of heritage to the community at both micro- and macro-levels, and should be used whenever a full account of the market and non-market benefits of heritage is required.

Cultural value

A thorough economic evaluation of the market and non-market benefits of an item of heritage will tell us a great deal about the cultural value of the item, because in general the more highly people value things for cultural reasons, the more they will be willing to pay for them. Nevertheless, it may not tell the whole story, because there are some aspects of cultural value that cannot realistically be rendered in monetary terms (Fig 19). There are at least three reasons why this is so.

Fig 19 Not all cultural heritage is old. The Sydney Opera House, designed by Danish architect Jørn Utzon, was completed in 1973. It has recently been nominated for inclusion in the World Heritage List. © David Throsby

First, some values are collective rather than individual, such that simply adding up individual valuations would fall short of a complete account. Think, for example, of the value of the music of Bach to the history of civilisation – it is a value that is likely to transcend the sum of individuals' willingness to pay. Second, there are some benefits to which individuals would find difficulty attaching a monetary value. An example might be national identity – it makes little sense to ask someone what it is worth to them in financial terms to be British, or French, or Australian. Third, some aspects of value may be intrinsic to the heritage item concerned; such values, if they exist, would be independent of individual willingness to pay, yet, if they are acknowledged, they would have a bearing on decision-making. Note that I use the term 'intrinsic' here in its usual sense of 'contained within' or 'inherent'; as such it differs somewhat from the use of 'intrinsic' in the 'cultural value triangle' in John Holden's and Robert Hewison's presentation at this conference. Similarly my use of the term 'cultural value' is confined to specifically cultural attributes, whereas theirs refers to the total value of cultural phenomena, including economic as well as cultural dimensions.

So although it may be claimed by a committed neoclassical economist that a full economic evaluation of the benefits of heritage is all that is necessary to capture the cultural significance of the heritage question, it is important to realise that other elements of value will remain unaccounted for. This brings us back to the concept of cultural value as I have referred to it above. We need to recognise that an independent assessment of cultural value will always be important in informing decision-making in regard to heritage, no matter how thorough an economic assessment is made.

The task ahead, then, is to devise systematic and replicable means for representing cultural value in ways that can assist in policy formulation. The difficulty here is one of measurement. The economic values are relatively easy to measure, at least in principle, because they can all ultimately be expressed in monetary terms. Cultural value, on the other hand, has no ready-made unit of account. The best we might do is to break cultural value up into its constituent elements as a means of assessing its dimensions. In the case of heritage these elements might include:

- *aesthetic value*: beauty, harmony
- *spiritual value*: understanding, enlightenment, insight
- *social value*: connection with others, a sense of identity
- *historical value*: connection with the past
- *symbolic value*: objects or sites as repositories or conveyors of meaning
- *authenticity value*: integrity, uniqueness.

These values derive from a broadly cultural discourse about the significance of art and culture in human affairs. It is obvious that cultural value in this context is multi-dimensional, qualitative, subjective and likely to change over time.

The way forward

Despite the difficulties of specifying and measuring cultural value, we need to develop procedures for putting into effect appropriate appraisal procedures for heritage projects where *both* an economic evaluation *and* an assessment of cultural value effects are carried out in parallel. As noted above, the measurable economic variables are clear enough, though the pitfalls in measuring them should not be underestimated. In this connection it is worth remembering that, in the absence of a full-scale assessment of these economic values for a particular project, it may be possible to infer some of them from results obtained for other projects by a process known as 'value transfer'; however, the applicability of these methods to cultural heritage is still subject to debate. On the cultural value side, work is proceeding in identifying a range of indicators which, when taken together, can be thought of as providing an overview of the cultural value yielded by a given project. These indicators would ideally represent a mix of public or community preferences and expert judgement, so that the best-informed basis can be provided for decision-making.

Note: Aspects of the economics of heritage are discussed in Hutter and Rizzo (1997) and Peacock (1998). A fuller account of the application of cultural capital and sustainability principles to heritage conservation can be found in Throsby (2001). Several examples of the application of economic evaluation methods to cultural heritage are presented in Navrud and Ready (2002).

Capturing the impact of museums on learning

Sue Wilkinson

Director of Learning, Access, Renaissance and Regions, Museums, Libraries and Archives Council

I have been asked to talk to this conference about capturing the impact of museums on learning. This is one critical element of the work that the Museums, Libraries and Archives Council (MLA) has been carrying out both nationally and regionally on impact evaluation. We are also developing a framework for capturing the social impact of our sector, and a number of pieces of research on its economic impact are now under way. One of these has just been completed – an economic valuation of Bolton MLA services funded by Bolton Metropolitan Council and MLA North West. Other work in this area has been undertaken by Archives, Libraries and Museums London and by the South West Museum, Library and Archive Council.

MLA and the nine regional agencies have four key strategic aims, to:

- increase and sustain participation
- put museums, archives and libraries at the heart of national, regional and local life
- establish a world-class and sustainable sector
- lead sector strategy and policy development.

Critical to all of these is understanding and articulating the value of our sector. Our research programme is shaped by establishing need; identifying how investment has changed the sector; how this change has improved outcomes for users of our services; and what needs to be done to build on this in the future.

Three years ago we began work on a best-practice framework, Inspiring Learning for All, which was designed to identify best practice in terms of stimulating and supporting learning. At the same time, we commissioned the Research Centre for Museums and Galleries at the University of Leicester to work with us to develop a methodology for capturing and analysing information from users that would allow us to assess whether we had been successful in supporting learning and what we needed to do to improve. The methodology that emerged from this study is in many ways very similar to the contingent valuation model that has been discussed by others at this conference. While we were aware that the 'holy grail' was to be able to prove causal links between the experience of using MLAs and educational attainment or skills development, we knew that most of the evidence currently available was more descriptive, personal, individual and anecdotal. It focused on people describing their own perceptions of the impact MLAs had had on their learning or their lives and the value they placed on it. It remained highly subjective but could be triangulated by talking to teachers or tutors or other family members (depending on whether the learning impact we were trying to capture was formal or informal learning), and by talking to the institutions that had developed the learning experiences.

As the research to develop and test our methodology progressed, it became clear that such perceptions are critically important. What children think they have learnt from a museum visit – and what their teachers or parents observe about the changes that have taken place as a result of that visit – allow us, when placed in a proper research context, to make important observations about the power MLAs have to create confidence, shape attitudes to learning and build knowledge and understanding. What I now want to do is to describe briefly some of the evidence we have acquired and to show how we are beginning to look at ways of triangulating this against independently acquired evidence of the children's educational attainment.

The research, piloting and testing we carried out over a three-year period has enabled us to identify the key characteristics of the learning which takes place in MLAs (Fig 20). This is what users value about MLA-based learning. We have now used these to shape the evaluation of a range of different programmes. For example, we have used them to evaluate the summer reading campaign (which runs in 98 per cent of public library authorities every summer and is designed to encourage children to read during the school vacation); adult basic skills programmes; and a number of schools programmes run as part of the strategic commissioning programme jointly funded by the Department for Culture, Media and Sport and the Department for Education and Skills and as part of the Renaissance in the Regions programme. Renaissance in the Regions is a programme of government investment in regional museums that is designed, among other things, to enhance their ability to work more effectively with visitors of all ages.

To illustrate the sorts of information we now have about the impact our sector has on learning, I shall draw specifically on the Renaissance-funded work. Over the past two years we have carried out two separate studies that together provide us with time-series data on a huge scale.

In 2003 we carried out a survey in 36 museums that involved 22,000 schoolchildren and 835 teachers. In 2005 we repeated this survey, using the same questions but this time running it in 69 museums with 1,643 teachers and 26,791 pupils. The data from the second survey are still being analysed but the results look enormously interesting and have confirmed for us that the Generic Learning Outcomes do capture and describe the type of learning that takes place in our institutions and the importance teachers attach to museum visits in supporting both teaching and learning.

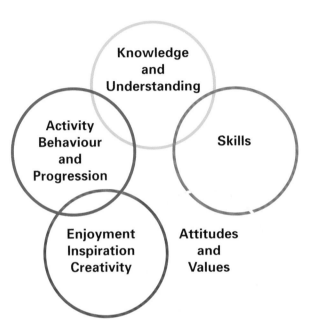

Fig 20 Generic Learning Outcomes.

The type of evidence that the Generic Learning Outcomes methodology enables us to describe is both quantitative and qualitative. It gives us a framework for questioning people about the value they place on the experiences that MLAs offer as well as a way of analysing their responses. It is a methodology that can be applied to both users and non-users of services because it is able not only to ask questions about impact but also to analyse people's perceptions of what MLAs can do and might do in the future.

When we asked teachers in the 2003 survey what they thought of museums they told us that they are:

    Places where children gain new knowledge and understanding (86%)
    Enjoyable places to learn (97%)
    Places to arouse curiosity (86%)
    Places of inspiration (64%)
    Places that promote tolerance and understanding of other cultures (44%).

In the same survey we also found that:
> 94% of teachers attending a museum activity see it linking directly to the national curriculum
> 94% of children aged 7–11 felt that they had learnt new things
> 58% of children aged 11–14 think a museum visit makes schoolwork more inspiring.

The data from the 2005 survey is still coming in but the emerging pattern is that it supports and reinforces the conclusions of 2003, but with some important additions that seem to demonstrate the power of investment in our sector:
> 92% of teachers think pupils will feel more positive about learning as an outcome of their museum visit
> 95% of teachers think their pupils will have acquired new subject-specific facts
> 86% of Key Stage 2 children thought museums were exciting places
> 83% of Key Stage 3 pupils thought museums were good places to learn in a different way from school.

Other important facts that we have learnt from a combination of the 2003 and 2005 surveys are:
> 40% increase in use of museums by schools since 2003
> 19% of visits are from schools in 10% of the most deprived electoral wards (2003)
> 38% of schools visiting Renaissance-funded museums have between 25% and 100% of their pupils eligible for free school meals (2005)
> 45% of teachers visiting museum 'hubs' are doing so for the first time (2003)
> 85% of teachers surveyed came from schools that made regular use of cultural institutions.

The research suggests that museums and schools are developing more integrated ways of working, that there has been a large increase in the number of contacts between schools and museums and that teachers feel significantly more confident about using museums to support their learning than they did two years earlier. The research also shows the relationship between the Generic Learning Outcomes; it seems that what museums are good at is creating an emotional engagement that inspires children to learn, and stimulates the acquisition of knowledge and skills.

All of this data is, of course, presented in the context of a huge amount of information about the types of schools that museums are working with. Government investment has focused on reaching new audiences who may not typically use our sectors, and the evidence of both the 2003 survey and the 2005 follow-up, which analysed both postcode data and data about free school meals, has shown that the Renaissance-funded museums are working with disproportionately large numbers of schools serving children from socially deprived circumstances.

I said earlier that the 'holy grail' in terms of the learning agenda is to be able to demonstrate a causal link between the MLA experience and attainment. We have been thinking about a way to capture this through looking at course-work results over a period of time and we tried this approach out with one of our case-study primary schools in the 2005 survey (Fig 21). The results are startling and we may well commission work to do more in-depth work of this nature over the course of the next few years.

Name	KS2 SAT English	KS3 History - no visit	KS3 History – no visit	KS3 History post visit
Trish	4	5	4/5	6/7
Tom	4	3/4	4/5	5
John	4	New to school	New to school	5
Matt	4	6	5/6	7/8
Cath	4	5/6	5/6	6/7
Oscar	4	5/6	6	7

Fig 21 Measuring the benefit of museums visits: course-work results from Downham Market School, Norfolk.

What I have just described, albeit very briefly, is our work on learning impacts. For the last 18 months we have been looking at ways of extending the Generic Learning Outcomes to capture social impacts, and in particular the impact that our work is having on communities as well as on individuals. In partnership with our regional agencies we have been using a social-capital methodology to find a way of establishing what the social outcomes of the work of MLAs might be, so that we can then look at the extent to which these are currently being delivered. The approach we are piloting and testing is intended to develop and extend the work of François Matarasso (1997). His research has talked about social value as an aggregated value for individuals. Our methodology maps

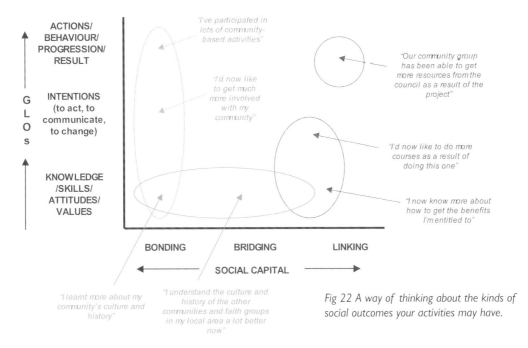

Fig 22 A way of thinking about the kinds of social outcomes your activities may have.

outcomes against activities and requires an analysis of the perceptions of individuals who have participated in programmes to be triangulated against the views of group leaders (teachers, community and faith leaders, etc) and of MLAs themselves (Fig 22).

It is this sort of analysis that we feel is putting our parallel work on economic impact into context. If, as our economic impact studies seem to show, users and non-users of our services put a very high financial value on them, then we have to ask ourselves why this is the case. The pilot work we have done on social and learning impacts is now providing some of the answers to these questions. People value our sector for a wide range of reasons. Not only do they feel that museums and libraries have a great deal to offer to their own learning and to their children's learning, they also think they can make a real difference to their lives. One of the comments in the Bolton survey of economic impact was that the local library was 'the lifeblood of the community' (Fig 23).

Fig 23 The results of enhanced learning programmes: happy and engaged children.
Jonathan Goldberg © MLA

Public space: public value

Julia Thrift
Director, CABE Space

This paper brings together work that has been done by colleagues from across CABE, looking at issues to do with the value of public space and the concept of public value as a whole, as applied to the built environment.

For those of you who are unfamiliar with CABE and CABE Space, it is worth outlining who we are and what we do. CABE, the Commission for Architecture and the Built Environment, is a non-departmental public body (ie a quango), which was set up by the Department for Culture, Media and Sport in 1999. CABE was established following Lord Rogers' Urban Task Force, which identified that if the government wanted more people to live in cities, then the quality of their architecture and urban design would have to be improved. So CABE was given a remit to champion improvements to the quality of England's architecture and public spaces.

Following the work of Lord Rogers, the Office of the Deputy Prime Minister set up another task force, the Urban Green Spaces Task Force, to examine why the country's urban parks had declined so very badly. It reported back to government in 2002 with a wide range of recommendations, one of which was for the establishment of a national organisation that could champion better parks and public spaces. As a result, the Office of the Deputy Prime Minister asked CABE to take on this role. Following discussions, it was agreed that a new unit would be set up within CABE, with a remit to champion improvements to the whole urban public realm, but concentrating initially on improving urban parks and green spaces. In May 2003, CABE Space was launched with funding from the Office of the Deputy Prime Minister but as part of an organisation that is sponsored by the Department for Culture, Media and Sport – which gives us good links between the two government departments.

Fig 24 Parks and squares: who cares? *Listening to what people think about their public spaces (CABE 2005b).*

Right from the outset, CABE has always been very interested in issues about value. One of our first publications was *The Value of Good Design* (CABE 2002), which made a strong case that spending money on design was an investment that reaped dividends and was not – as so often perceived – an additional cost. Since then, arguments around value have become more sophisticated but the issue of value remains key to CABE's work. We were delighted when, in 2003, the Treasury's 'Green Book' recognised that good design has both a social and economic value.

When CABE Space was launched in 2003, one of the first things we published was *The Value of Public Space* (CABE 2004a). This brought together a wide range of information and research to support the argument that high-quality public space has a range of benefits. It collected the information in terms of themes: the economic benefits, the physical and mental health benefits, the benefits for biodiversity, etc.

We followed this up with two more substantial pieces of new research. The first, published as *Is the Grass Greener?* (CABE 2004b), looks at 11 cities around the world that have taken a long-term strategic approach to improving public space and examines the value that they have gained from it. The benefits of having networks of high-quality, well-designed, and well-maintained public spaces include: an enhanced reputation among investors and tourists, social and health benefits, etc. Many of these cities – all of which are very different in terms of location and economy – are frequently cited in lists of great places to live.

A further piece of research, published as *Does Money Grow on Trees?* (CABE 2005a), captures the economic benefits of high-quality parks and green spaces. Put simply, if you invest in improving a park, does this do anything to improve the local economy? Tracking the relation proved difficult, but the answer was 'yes'. Some of the most telling information that the research revealed was the way that key decision-makers spoke about places – for instance, 'We decided to put our new building here because of the lovely park', and that sort of thing.

What I have talked about so far is our attempt to capture the economic value of high-quality, well-designed and well-maintained buildings and public spaces. But there are other sorts of value that are, perhaps, less quantifiable but often equally important – for instance, the value that we, as people, attach to places.

In 2004, CABE Space published a *Manifesto for Better Public Spaces* (CABE 2004c). We set out 10 reasons why we thought public space is so important and why investing in it should be a high priority. We invited the public to respond. Did they agree? And what did they think about their local parks and public spaces?

We had a fantastic response and received 1,500 individual comments from people telling us about how much they valued their spaces (CABE 2005b; Fig 24). We hit a nerve: we discovered that people really are passionate about the quality of their local public space. This was borne out by research by MORI that discovered that 91 per cent of the public thinks that good parks and public spaces improve their quality of life (Fig 25). This is an amazing finding: I cannot think of anything else that nine out of ten people agree about.

So it is clear what people think about public space in general – but we need a way of capturing what people value about individual sites and places. In order to be able to do this, CABE Space has been developing what is currently known as (and this is still a working

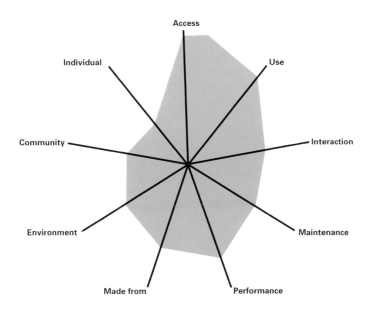

Fig 25 The CABE 'Place Consultation Tool' uses spider-diagrams to compare the way in which different groups of users perceive the significance and value of local open spaces. © CABE

title) the Place Consultation Tool. This tool is based on something called the Design Quality Indicator, which was developed by the Construction Industry Council, with support from CABE. The Design Quality Indicator is a way of capturing what people think about an individual building – is it well designed or not? It is easy to use but is based on a huge amount of research; it is a tool that carefully weights different considerations, using some highly sophisticated equations, to come up with a simple assessment of what is good about the building and what is not. The Design Quality Indicator took many years to develop, and, rather than starting from scratch, CABE Space and the Construction Industry Council took this work as the basis for developing the Place Consultation Tool, which we are currently piloting and refining.

The Place Consultation Tool is an easy-to-use tool that assesses what people think of a particular public space. It can be used to evaluate all types of public space – parks, streets, etc. You can use it in many different ways. For instance, you can use it to compare what different types of people think of a particular space: what do old people think of it? And how does this compare with what teenagers think of it? You can use it to track changes in people's perceptions over time: for instance, what do they think before work is carried out, and after?

One of the pilot studies looked at a particular park and compared what the park managers thought about it with what the local community group felt about it. The park managers thought that the park really needed to have better standards of horticulture – but they thought that the community consultation they did was really quite good. In contrast, the community group thought the horticulture was fine – but they thought the community consultation was poor.

Having discovered differences like this, it is then possible to negotiate how the service delivering in that space can be improved and how money can be focused on providing what local people want, not just what professionals think they want. So the Place Consultation Tool begins to give us a methodology for capturing some of the value that the public perceives local places to have.

Fig 26 Funding long-term maintenance remains an intractable problem. © CABE

I would now like to move on to some other work that CABE is doing. Again, it is work in progress. We are very interested in the value of the built environment – something that we call 'physical capital'. In 2005 we published a selection of essays by leading thinkers to help stimulate debate around this emerging topic (CABE 2005c).

One of the ideas that came out of this was the need to look at different ways of capturing value in terms of physical capital and the methodologies that could be applied. We are taking this forward with a piece of research on value mapping, which will be published later in 2006.

In addition, we will soon be publishing a very practical guidance for decision-makers that brings together relevant information about value-mapping methodologies in a simple and easy-to-use format. The aim of this is to make some of the more academic and obscure information about value more accessible to decision-makers who need quick and easy access to the information. This, too, will be published later in 2006.

Finally, as part of our work exploring the idea of physical capital, we will be publishing a book of essays on the subject of 'The Cost of Bad Design'. It will comprise five thought-provoking essays that will explore the idea that, as a society, we are paying for past design mistakes in terms of poor buildings and public spaces. Good design is often – and unfairly – perceived to be expensive. But how much are we paying for the consequences of bad design?

I would now like to raise another issue that we in CABE Space are very much aware of: the cost of poor maintenance. When the Urban Green Spaces Task Force was exploring why our public parks had declined into such a dire state by the end of the 1990s, it concluded that one of the main reasons was that the money for maintaining them had been gradually cut, year after year. The result of this gradual but ongoing budget-cutting is that high-quality parks, full of beautiful plants, cafés, paddling pools and other community facilities, have ended up as semi-derelict sites, with very little value to their communities.

This is not just an issue about parks, however. In this country we do not take maintenance very seriously – we create things, then let them decline due to a lack of revenue funding to pay for cleaning and general maintenance. Once the asset has declined into a truly awful state, we then spend a huge amount of capital rebuilding it. This is exactly what has happened in parks and public spaces all over the country. We cannot let this go on, however. It is surely a huge waste of public money.

We are now in the middle of the largest public building boom that we have seen for a generation, with some £38 billion being spent on delivering the Sustainable Communities plan alone. If we look back to the places we created in the last major public building boom – in the 1960s and early 1970s – what do we find? I would argue that many of the places created then did not have enough money spent on their public spaces in the first place, and also, once those places were created, they were not maintained. The tower blocks of the 1960s often had relatively spacious flats with great views, but the spaces around them are now derelict wastelands (Fig 26). Who would want to point to a place like that and say with pride, 'I live there'? CABE, with many others, is working to try to ensure that the places we create through the Sustainable Communities plan are as good as the high aspirations set out in the policy documents – well-designed places, that people really want to live in. (Fig 27)

However, there is a real risk that these places will not be maintained (to ensure that they remain) as beautiful, desirable places in which to live. Local authorities struggle to find enough money to spend on this sort of dull but essential activity. Many are now taking this issue very seriously, and huge improvements have been made, but there is still a risk of not being able to look after these places to a high enough standard to ensure that they remain attractive and high-quality places.

I do not have a solution to this problem, but I do have a suggestion. One of the reasons that local authorities often do not really value public space as highly as they might is that public space does not feature highly on their registers of assets. As Geoff Mulgan pointed out in his essay for our *Physical Capital* publication (CABE 2005c), the value of public spaces 'is generally taken to be the value of alternative possible uses ... but also reflects the legal status of the land'.

Fig 27 Maintaining space matters: will today's sustainable communities be tomorrow's candidates for demolition? © CABE

So, applying this criterion to an ordinary public park, what do we get? The park is likely to be designated as public open space. Therefore, it is illegal to use it as anything else – so it has no development value, no 'alternative use' value. Therefore, it is given zero value on the local authority's register of assets.

Compare this with a museum building: it is possible to get a commercial estate agent round to look at the building and say how much it would be worth if it were sold. This value can then be put on the register of assets, and the accountants will then expect a certain percentage of this value to be spent on maintaining it.

So, coming back to our public park – let us think of its value in another way. What if it were flattened by a bomb and we had to re-create it (Fig 28)? Well, we would then have to purchase:

- topsoil (this is very valuable and expensive)
- trees (a mature tree costs about £5,000 from a nursery)
- railings (could easily cost £1 million)
- cafés and other buildings
- landscape design
- earthworks
- shrubs
- paths (very expensive)
- signage

And so on. Even a very ordinary park would cost many millions if valued in this way. There is, therefore, the possibility of using this argument to make it clear to local authorities that their public parks are hugely valuable assets that represent a major investment of public money, and should be looked after accordingly.

So, in conclusion:

- we know from our research that 91 per cent of the public thinks that public space is important
- research shows that high-quality public space adds value to places
- not maintaining public space wastes public money
- we should consider changing the way that non-developable public space is valued in registers of assets.

Fig 28 What is the value of a typical park? © CABE

Heritage on the front line: the role of a heritage champion in North Yorkshire

Heather Garnett

County Councillor and Heritage Champion, North Yorkshire County Council

The purpose of this paper is to focus on the sharp end, the delivery of heritage objectives, exploring the topic from the perspective of a heritage champion working within a local authority, North Yorkshire County Council. It demonstrates how the historic environment plays a key role in the sustainable economic and social success of the Yorkshire and Humber region and how the role of a heritage champion adds value to this process.

North Yorkshire is England's largest county. It is richly endowed with a stunning built heritage, breathtaking landscapes and diverse social histories that give it a unique regional role in generating wealth, health and inward investment. However, in order to maintain the benefits, it is essential that we continue to offer a quality product. In short, the sustained economic success of the county depends on ongoing support for North Yorkshire's heritage resources. The role of heritage champion makes a strong contribution to this objective.

The county is endowed with more than 12,200 listed buildings, 1,700 Scheduled Ancient Monuments and a World Heritage Site. With such assets and a largely rural economy, heritage schemes in North Yorkshire operate at the centre of local communities. The benefits heritage brings to the county can be summed up in three key points:

- building communities and encouraging respect
- a quality product equals economic success
- helping to secure a sustainable future.

To consider the first point: heritage is not just about buildings and places but also about people and bringing people together. The list below gives a few examples from the many that are happening around the county of how a commitment to heritage involves, enthuses and brings to life the relevance of history to create community cohesion and a true sense of belonging. This is essential in today's world of fragmented families and communities.

Secondly, without an outstanding product, outstanding success cannot be achieved. A heritage champion is tasked with promoting the county's environmental and social heritage assets at multiple levels to multiple stakeholders so that the right decisions are made to use these assets to best effect.

Thirdly the environment is changing – for example, global warming and the poor economics of upland farming are having a profound effect on life in North Yorkshire. Appropriate support for the historic environment is central to creating sustainable solutions when finding new uses for redundant buildings, developing highways schemes with minimum adverse environmental impact and managing tourism pressures while simultaneously supporting communities and increasing accessibility. Balancing these diverse, often seemingly conflicting tasks is complex but caring for the built and natural heritage is central to a sustainable future.

The examples below have been chosen to demonstrate the three critical points: the way heritage contributes to building communities; economic success through creating a quality product; and finding sustainable solutions. They range from local community initiatives to projects of international importance:

- Thornborough Henges
- City of Ripon regeneration scheme
- The Discovery Centre
- Hawnby floods
- Nidderdale Area of Outstanding Natural Beauty.

'Looking after our heritage and our environment – in our countryside and in our towns and villages' is one of North Yorkshire County Council's key objectives. It goes right to the heart of local communities as does policy relating to the delivery of its services, which includes retaining the county's little village schools, maintaining its minor roads and keeping its mobile libraries out and about. Neither should we underestimate the benefits that heritage projects, keying into the national curriculum, bring to education. Throughout the county children are becoming involved. Understanding the past brings about a clearer perception of their own lives and prepares them for the future.

It is within this context that more than 200 active community-based heritage projects operate. While some are major projects many others are seemingly small, but when taken together they contribute fundamentally to the well-being of the county's population and to its infrastructure. The examples selected here are just the tip of the iceberg.

The first example, the Scheduled Ancient Monument, Thornborough Henges, was originally a local heritage interest but has developed into a significant national and international concern because of perceived threats to its survival from aggregate extraction (Fig 29). Although seen as a 'villain of the piece' when wearing its Mineral Planning Authority hat, the county council has now successfully brought together landowners, the quarrying company, English Heritage, parish and district councils and pressure groups and set up the Thornborough Henges Working Party to agree a management and interpretation strategy for the site. As identified in the Sub-Regional Investment Plan, North Yorkshire County Council needs to develop new tourist attractions to enable it to maintain and increase its market share in the tourist industry. The centrally located henges, near to the A1, have the potential to be a prime attractor. The heritage champion's role here is to foster clear thinking and to find sustainable solutions within the county council, a complex area in which strong tensions exist between the various responsibilities of development, mineral extraction, asset management, educational opportunities and tourism.

The second example is a project that contributes to all three key points of building communities, economic success and sustainability. The City of Ripon Market Town Regeneration Project is a beacon of success, exceeding nearly all its social and economic targets by a large margin. Ripon is the UK's smallest cathedral city, with a population of just 16,500, and it is one of the most ancient – its charter was granted

Fig 29 An aerial view of the three early Bronze Age Thornborough henge monuments, without which no discussion of heritage issues in North Yorkshire in 2006 would be complete. This image was taken during the filming of the popular Time Flyers *archaeology TV programme.*
© *Dave Macleod, English Heritage*

by Alfred the Great in 886. However, over the years its importance both commercially and as a service centre meeting the needs of people from surrounding villages has diminished and urgent action was needed to reverse the cycle of decline. At the start of this project, earnings were 60 per cent of the national average in Ripon's rural hinterland, and within the city itself, 80 per cent of households were unable to afford market housing. But Ripon had many things in its favour, not least a rich history and the unrealised potential of its many historic buildings. In essence, Ripon had the raw materials to develop its role as a valuable service centre, to attract tourism and to thrive. The figures in Table 1, showing Ripon's success, speak for themselves.

Table 1. Ripon's success story

	Target	Achieved
Buildings improved	14	75
Community facilities	9	59
New businesses	16	64
Jobs created	55	168
New voluntary workers	368	620
People trained	46	4,104

The make-over included an upgrade of Ripon's 12th-century market place (Fig 30): it was re-cobbled, listed buildings were restored and a public space was included for community activities, providing a venue for the horn-blowing 'setting of the watch' ceremony that is still undertaken at 9 pm every night by the town's Wakeman, a figure who for

Fig 30 Market day in the city of Ripon before the re-cobbling of the market square. The foreground, in front of the obelisk, is now a public open space. © North Yorkshire County Council

centuries had charge of the town and its inhabitants' safety. Membership of the Partnership Executive provided opportunities for the heritage champion to contribute to the development, execution and assessment of the regeneration project and to liaise with the county council to encourage projects such as the sale of the listed redundant 19th-century workhouse hospital to the Ripon Council for Voluntary Service. Now sensitively restored as Community House, the former hospital is a welcoming hub that offers meeting, counselling and learning facilities, a children's centre, community transport facilities, a café and gardens.

A project at an early stage of development, of sub-regional and regional significance, is the Discovery Centre, the third example. Creation of this state-of-the-art family and local

Fig 31 Floods in the North Yorkshire Moors National Park washed away bridges, isolating Hawnby village; steep single-tracked roads made it difficult to install temporary bridges. © North Yorkshire County Council

history centre in Harrogate, the county's largest town, in order to serve the large and growing public interest in family history, is a major inclusive project due for completion in 2009 dependent upon funding support not yet obtained. The experience of Jeremy Paxton, whose mother was born in Yorkshire, displaying his emotions publicly in the recent BBC 2 television programme, *Who do you think you are?*, mirrors that of many people discovering their own family histories. The centre will use modern techniques, including digital photography and microfilming, to bring together copies of records that can be used for family and local history. It will offer friendly first-class research facilities, education and display facilities, a shop, cafeteria and lecture theatre. It will attract people in search of their roots from across the world and develop links with the tourist trade on a county-wide basis. This sustainable solution offers both widespread accessibility to many of the county's most popular records and at the same time protection of the original fragile archive material from damage through over-handling.

The fourth example is a small community in the North Yorkshire Moors National Park. Here, in June 2005, people's lives were turned upside down by a flood that cut off the village of Hawnby and other nearby villages (Fig 31). Homes and businesses were destroyed or inundated. One man nearly drowned when rescuing a dog from his boarding kennels and was plucked from his roof and transported by helicopter to safety. Three of the county's listed bridges were washed away cutting off all access to the village and other homes. It was vital, in a county still recovering from Foot and Mouth Disease, to get temporary bridges in place as soon as possible, to get people and businesses on the move and to tell the world that North Yorkshire was open for business! To provide a sense of scale: North Yorkshire has 1,958 road bridges, 78 per cent of which are more than 150 years old, 80 of which are Scheduled Ancient Monuments and 150 of which are listed. A sustainable whole-life maintenance regime is in the process of being implemented, amounting to an annual budget of more than £4 million, which is set to reduce over time. The heritage champion took a leading role in the development and adoption of a sustainability policy for the county council. Over time the effects of this policy, as it is implemented, will impact on almost all of the authority's services.

My final example is Nidderdale Area of Outstanding Natural Beauty (AONB). AONBs offer many opportunities for public engagement with the natural and built environment (Fig 32). Nidderdale AONB teems with community activity given direction by the lead organisation, Harrogate Borough Council, working closely with its partners, which include North Yorkshire County Council, local people and government agencies. Expenditure in 2004/5 was £244,635 to which the county council contributed £41,197 demonstrating the leverage obtained from a comparatively small investment. With it much has been achieved, including since June 2002 the following:

- 63 new projects have been funded, varying from wildlife to the restoration of historically important buildings
- 10,000+ hours of volunteer time have turned helpers into tree wardens, dry-stone wallers, open-access volunteers (Countryside and Rights of Way Act) and countryside volunteers
- 150km of public rights of way have been improved

- 50 young rangers have been recruited
- 5,600 hours of adult education have been provided on countryside issues
- about 1,000 people have enjoyed events organised by the AONB
- 75 planning applications have been commented on
- in its first 12 months, the Friends of Nidderdale AONB has recruited 300 members
- And there is much more …

A dozen multi-disciplinary ecology, social history and community archaeology projects reach out with elemental force drawing people together to unearth the secrets of the past, creating for them a sense of identity and belonging. They find, for instance, evidence of Nidderdale's long industrial heritage including at least eight centuries of organised lead-mining. Nowadays hikers tramp the hills on tracks once trodden by packhorses transporting lead pigs to the east-coast ports for shipping to the continent – Yorkshire lead adorns the roof of Cologne Cathedral, for instance. Access to this history is provided by a heritage trail; the humps, bumps, holes in the ground and bits of surviving masonry of this working landscape are given meaning by guided walks and interpretation panels.

Fig 32 These walls in Dacre parish, within the Nidderdale Area of Outstanding Beauty, survive from the mid-19th-century Enclosures Act and form a distinctive feature of the landscape. © Nidderdale Area of Outstanding Natural Beauty

Quarrying in the AONB, as in other parts of the county, continues to provide employment and to mould the landscape, as it has for centuries past. Careful mitigation schemes are put in place through the planning process. There is a perception of an enduring, unchanging countryside but that notion is belied by the evidence on the ground: these valleys have adjusted to many changes over the centuries, as they continue to do. The landscape has been fashioned by man. The task now is to ensure that change benefits the landscape and brings prosperity to the AONB's rural communities. The following are some of the groups operating within the Nidderdale AONB: the Joint Advisory Committee; the Planning and Sustainable Development Fund sub-group; the Green Lanes Liaison Group; local heritage projects; the Friends of Nidderdale AONB; parish councils; the Pateley Bridge Quarry Liaison Group; and Nidderdale Plus. Involvement with all these bodies offers opportunities for a heritage champion to get involved at a local level and to foster good communication and a close working relationship between the partnership organisations, stakeholders, user groups and individuals. Both the formal and informal channels of local government offer similar pathways at a more strategic level.

These examples have brought to life the role of a heritage champion, whose tasks are summarised in Table 2.

Table 2. The role of the heritage champion

Overview	Day-to-day
Leadership	Facilitate
Public face	Encourage
Ambassador	Promote
	Question/challenge
	Spread, gather and develop ideas

The question now is, 'Where to next: how can the role of heritage champion be taken forward?' Within North Yorkshire County Council there is a need to develop the leadership role within the authority in order to create better inter-departmental links and integrated thinking and to develop policy. Looking outside the authority, a key area to strengthen is that of the council's relationship with external partners and stakeholders and with the private sector both within the sub-region and at regional level to help achieve shared objectives and reach sustainable solutions. A further objective is to focus more on community, education and outreach projects to reach a broader section of society within and without the county. Moreover, it is essential to continue to uphold the public value of heritage to ensure that the contribution it makes to the economy and the community's quality of life is fully understood and supported. Attendance at, and taking part in conferences such as the one today provides opportunities to do this at national level, and to network, share experiences and gain insight and perspectives into issues on a broader front. Furthermore, it puts North Yorkshire on the map.

To conclude: the role of a heritage champion is far, far more than promoting the well-known picturesque beauty of Yorkshire as encapsulated in such populist images of the county as Kilnsey Crag, in the heart of *Calendar Girls* country. The role is absolutely rooted in today's challenges of creating community cohesion, a sustainable future and economic success in a county that has diverse and very real needs.

Fig 33 In 2005, to coincide with the Castleford Heritage Festival and the Regeneration Project, the Castleford Schools Pyramid for Arts (CASPA) commissioned local composer and teacher David Hookham to write some songs specifically about Castleford and its heritage. The result was *Legeolium Live*, a series of 13 songs tracing Castleford's past and present. The work was previewed last July at the Heritage Festival in Castleford. Members of the choir came down to London specially to perform four of their songs at the conference. © Heritage Lottery Fund

From significance to sustainability

Kate Clark
Deputy Director, Policy and Research, Heritage Lottery Fund

This session has really been about the instrumental values of heritage – if you like the benefits that looking after heritage brings. This distinction between instrumental and 'intrinsic' values is, I think, a really useful and indeed powerful one.

The idea of 'intrinsic' values in heritage is nothing new – indeed what makes something part of the heritage is the fact that it is valued. Anyone who has ever managed a heritage site, fought to prevent the demolition of a much-loved building, or applied for lottery funding has made that case on the basis of an argument about values. Indeed, if you do not understand what is important, how can you possibly make decisions about it?

When we protect a building or site, it is done because it is of value to the public. The whole justification for regulation, intervention or subsidy in heritage is based on the idea that these assets are important not just to us as individuals, but to the wider community. Yet we rarely think about how we define that value – and whose values we take into account. A lot of the traditional charters and conservation philosophies seem to assume – as William Morris did – that these are things over which 'every educated gentleman would agree'. The listed building legislation is based on the concept of 'special architectural and historic interest', but we rarely discuss what those mean and how we interpret them. The big conservation questions have been about whether or not we should restore something – not whether it mattered or not.

Significance

The lid came off Pandora's box in 1988 in a small mining town somewhere in South Australia. A group of Australian heritage experts got together to draft the Burra Charter (ICOMOS 1988, and see also http://www.icomos.org/Australia). For the first time, the idea of significance at the heart of heritage was out in the open. Since then – like it or not – it has been hard to ignore the issue of what matters to whom and why in heritage – in other words, the so-called 'intrinsic' values of heritage.

Nearly ten years ago we had a conference about values in heritage – then it was because English Heritage, the National Trust and the Heritage Lottery Fund were working together to introduce a new approach to looking after heritage that put value at the heart of decision-making. We introduced the idea of conservation plans – which were all about the need to understand what was important about a place before making important decisions (Clark 1999; Heritage Lottery Fund 2005). Since then I have seen plans for everything from a tiny pocket of land to the British Museum.

At the time, we were warned that we were creating something bureaucratic and unnecessary, and that talking about values simply diverted funding from essential repairs. And there is no doubt that some of the planning that has gone on has not been worthwhile. But the bigger problem is that understanding values is more difficult than we first thought. It is one thing to be an expert architectural historian or naturalist for example;

working with a community group to identify what is important to them and then translating those values into action without losing sight of the core issue of stewardship is very different. Values-based planning was controversial then – and still is difficult – but ultimately, it is what heritage is all about.

Sustainability

But it is one thing to repair or conserve a heritage site; it is all together another to find a future for it. One of the big changes in environmental philosophy over the past 20 years has been the idea of sustainable development – development that meets the needs of present generations without compromising the ability of future generations to meet their needs. Sustainable development brought together conservation and wider economic and social agendas. And if we were going to engage with those agendas we needed data, particularly on the social and economic – or instrumental – benefits of heritage.

These were very different to the kinds of values we were used to dealing with. Conservation planning forced us to look at why a site mattered to people – this was about what benefits would be delivered by looking after that site. We needed economists and social scientists, not archaeologists or curators. And it was this other kind of value that we have been grappling with at the Heritage Lottery Fund. It is a relatively straightforward matter to account for where our £3.3 billion has gone and how many grants we have awarded – it is much more difficult to find out what that funding has achieved or what benefits it had delivered. In the last few years we have come a long way. We have made much better use of our own data in order to report on outputs – we can count education posts and conserved buildings and hectares of land conserved. And under Gareth Maeer, an environmental economist, we have put in place a programme of impact research, looking at the wider social, economic and environmental impacts of what we have done.

At the same time, the rest of the sector is also producing new evidence for the instrumental benefits of heritage; the annual *Heritage Counts* report in England is an excellent review of emerging data; as we have seen at this conference, the Museums, Libraries and Archives Council, English Heritage, the National Trust, and others are all producing important research on the impact and benefits of heritage.

The move towards sustainable development and thinking about instrumental benefits has forced all of us in conservation to engage much more with economics – the impact of what we did on the wider national economy, and the economic value of heritage for society. Suddenly we needed to be able to understand contingent valuation and the travel-cost method. We found ourselves thinking in a different way – but we were also worried. Was it really possible to reduce the value of heritage to a crude multiplier or the sum a member of the public would be willing to pay? Here were two different worlds with two different languages, both talking about ideas of value and both (secretly) convinced that the other was both unscientific and dangerous.

This session of the conference came out of those concerns. A group of us calling ourselves UKHERG – the UK Heritage Research Group – had got together to try to co-ordinate policy research in heritage and to share some of our worries (http://www.heritagelink.org.uk/sector.asp). Many of us were in the same position – experts in our field but without either the knowledge or resources to take on this new kind of research. It would have taken a whole conference to deal with all the new evidence that is emerging. What we wanted to do instead was to take a high level overview of the issues and also to remind ourselves that the instrumental benefits of what we are do are important – but that they are only part of the story. The person who has done most to reassure us that it is possible to bridge the divide between economic ideas of value and the kinds of values or significance that we deal with in heritage, is David Throsby, and we were delighted that he was able to be at the conference.

Discussion 2. The instrumental benefits of heritage: how are they measured?

Chair: Kate Clark (Heritage Lottery Fund)

Panel

Professor Randall Mason (School of Design, University of Pennsylvania)
Ece Ozdemiroglu (Economics for the Environment Consultancy)
Professor David Throsby (Macquarie University)
Sue Wilkinson (Museums, Libraries and Archives Council)

The Issues

- How do we persuade economists to look beyond the cheapest option and towards some of the wider instrumental benefits of the heritage?
- How does our progress in the UK compare with that in the USA and Australia?
- How should we move on from measuring the instrumental benefits of heritage to individuals to their value for the whole community?

The words below are not necessarily exactly those that were spoken, and are intended instead to capture the overall flavour of the discussion.

Ece Ozdemiroglu

Economists *do* look beyond the cheapest option; they *are* interested in the best, and they *do* recognise that people's values are important. In the past, the main environmental drivers to economic valuation have been regulation (ie costs/benefits of action/inaction) and the guidelines of the Treasury Green Book (ie the need to quantify as much as possible). Over the last decade there have been literally thousands of environmental valuations, but only a few dozen cultural ones.

Chair: We've heard a bit about Australia; what has been happening in this field in the US?
Randall Mason

Not as much as in the UK, and it is very good that you *are* taking a lead in its application. And it is not a matter of one approach being better than the other; in reality we need both qualitative *and* quantitative methods. It is important to remember that the public is not a simple entity, but something much more complicated and diverse: we therefore need to be very fluent and flexible in the way we talk to it. The concept of cultural capital may be the way to build a bridge between economists and heritage people.

Chair: The Australian government's Productivity Commission has just published draft proposals suggesting that the statutory designation of heritage assets should depend on financial compensation for their owners. As an economist, David, what do you think of this?
David Throsby

Even from the standpoint of an economist this appears to be an unhelpfully blunt market solution to the problem of private owners facing constraints of designation. It flies in the face of the idea of public value because its logic is that preservation can only be justified for pure economic reasons.

Chair: Sue – isn't one of the biggest challenges in looking at social impact the question of how we move from measuring individual benefits to looking at benefits for the community as a whole?

Sue Wilkinson

It is very challenging, but there are ways in which it can be achieved by aggregating the feedback from lots of different members of the public. Another useful approach is to ask people their perceptions not only as individuals but as members of the community, and then in turn to seek the independent views of acknowledged community leaders. Above all, it is a process that needs to happen incrementally rather than in terms of big abrupt steps.

Delegate

The recent experience of taking a major heritage project through the Green Book process showed that Treasury economists still find it very hard to understand our priorities and qualitative arguments. Either we solve this communication problem, or we miss out.

Ece Ozdemiroglu

It is true that the economists have to be given an asset valuation they can understand: it is something that you simply cannot avoid.

Randall Mason

In the US you have to *prove* that preservation pays: that to protect something will be more profitable than neglecting it. But you have also to realise that in the case of the heritage vs Walmart, the heritage will *never* win.

Delegate

When looking for evidence with which to persuade government we should not forget that there is 40 years-worth of market research already in existence. The problem is that we are not exploiting it. It will only become affordable if we are prepared to co-operate.

Delegate

Poorer communities find it more difficult to bid for heritage resources than rich ones. If the principles of public value are adopted, will it help persuade corporate sponsors to switch their investment away from elite areas and towards those in greatest social need?

Sue Wilkinson

There is a real interest in new ways of levering corporate funding and this is something that we all need to be thinking about.

Delegate

How can individuals who voluntarily take responsibility for items of cultural capital (eg barn conversions) be supported for adding new public value?

David Throsby

This takes us back to the Australian productivity debate about compensation, to which there is no clear answer.

Ece Ozdemiroglu

In a recent study of the Lake District supported by English Heritage and Defra it has been shown that landscape character can be given a robust economic value, and that caring for historic farm buildings can put money back into the local economy. I would also like to mention a report on *Valuation of the Historic Environment* that we prepared last year for English Heritage, the Heritage Lottery Fund, the Department for Culture, Media and Sport and the Department for Transport (available at www.english-heritage.org.uk/valuation).

3 The Intrinsic Values of Heritage

David Lammy, MP
Minister for Culture, Department for Culture, Media and Sport
Community, identity and heritage

Sir Neil Cossons
Chairman, English Heritage
Capturing the value of places

Christina Cameron
Canada Research Chair in Built Heritage, University of Montreal
Value and integrity in cultural and natural heritage – from Parks Canada to World Heritage

Edward Impey
Director of Research and Standards, English Heritage
Why do places matter?
The new English Heritage conservation principles

Dame Liz Forgan
Chair, Heritage Lottery Fund
Capturing the opinions of people

Deborah Mattinson
Joint Chief Executive, Opinion Leader Research
The value of heritage: what does the public think?

Community, identity and heritage

David Lammy, MP
Minister for Culture, Department for Culture, Media and Sport

Today we are looking at practical issues. Here, I want to highlight some of the challenges and set the scene for the discussion on 'Whose values matter?' Let me begin by reflecting a little on the past, on events that have brought us, as a sector, to this conference today.

Since I have been Minister of Culture, I have been contemplating how and when the concept of heritage emerged. At what point did we, as a nation, start thinking about what it is that we value and start talking about it as our 'heritage'? Arguably, we can trace it back to another period of rapid economic and social change. We know that in the late-Victorian era, 'progress' was the watchword. The demands of urbanisation and industrialisation often seemed to have little regard for the civic fabric and while Victorian developers might have designed in the Gothic manner, they also merrily gutted much of our medieval heritage. In response, William Morris established the Society for the Protection of Ancient Buildings in 1877 both to preserve existing structures and to 'counteract the highly destructive "restoration" of medieval buildings'. Thus was born the heritage protection system that we have in place today, and that we are in the processes of streamlining and reforming.

Morris did not act alone. John Ruskin's other great disciple, Octavia Hill, went on to found the National Trust (Fig 35). Others joined the blossoming Arts and Crafts movement. All shared a concern for conserving the natural and built heritage from the worst excesses of urban, industrial society. Yes, it was a little elitist in conception, but none the less progressive in results: their vision was a radical desire for the British people to enjoy their national heritage – as Octavia Hill put it, to 'make lives noble, homes happy and family life good'.

So the heritage movement that emerged in the late 19th and early 20th centuries was driven to a large extent by a desire to reflect and promote social change, supported by an interest in curtailing the excesses of the new breed of designers and architects. It was the same impulse that in 1926 led Patrick Abercrombie

Fig 35 Alfriston Clergy House, Sussex. This 14th-century timber-framed thatched hall was acquired in 1896 by the newly founded National Trust. © NTPL/Andrew Butler

Fig 34 Capturing the values of a nation. Designed by George Gilbert Scott to commemorate the life of Queen Victoria's consort and simultaneously to celebrate the cultural achievements of 19th-century Britain, The Albert Memorial remains to this day a potent symbol of national identity. Nigel Corrie © English Heritage

and others to found the Council for the Preservation of Rural England, not as a wish to mummify the countryside, but to preserve a public connection with the past threatened by tearing development and social change. So, too, in the 1940s: as Liz Forgan reminded us last year, it was a Labour government committed to opening up the heritage for all that established the National Land Fund. Its remit was to buy areas of countryside, together with historic buildings, that would be opened up to the public as a memorial to those who had died in both world wars, preserving the physical aspects of nationhood that they had given their lives for. It was an act of living memorial never intended to be of monuments and medals, but of space, belonging to the people. It is worth recalling Chancellor Hugh Dalton's great words: 'It is surely fitting in this proud moment of history, when we are celebrating victory and deliverance from overwhelming evils and horrors, that we should make through this fund a thank-offering for victory, and a war memorial which many would think finer than any work of art in stone or bronze.' Here was a supremely inclusive idea of heritage – which today lives on in the National Heritage Memorial Fund.

But as the post-war consumer age took hold, the fund was sidelined. Britain, it was felt, had too much heritage. Extraordinarily, in 1957, the Treasury Minister and English patriot, Enoch Powell, dismantled the fund by slashing its budget by 80 per cent. The following decades witnessed a steady deterioration in our care for the natural and built environment. Our great medieval and Victorian city centres were gutted by planners and politicians, while our stock of historic houses was left to rot. It was the sale of Mentmore House and its contents in 1977 that kindled a public outcry and led to the creation of the National Heritage Memorial Fund, and with it – for the first time – the concept of 'national heritage' (Fig 36). Shortly afterwards, English Heritage emerged and the heritage world started looking much as we know it today.

Much of what flowed from this period of policy introspection was for the good. New safeguards, tax breaks and funds emerged, but at the same time, in certain circles,

Fig 36 The sale of Mentmore House and its contents in 1977 kindled a public outcry and led to the creation of the National Heritage Memorial Fund. © Crown copyright. NMR

Fig 37 Thanks to a £23,000 grant from the Heritage Lottery Fund, 30 youngsters were able to embark on a once-in-a-lifetime Young Roots project that saw them researching, interviewing and recording the history of football and the highs and lows of Manchester City Football Club. Joan Russell © Heritage Lottery Fund

heritage began to be a dirty word. Against the backdrop of ITV's *Brideshead Revisited*, heritage began to have inherently conservative, narrow-minded connotations. The poet Tom Paulin voiced it most succinctly when he said, 'the British heritage industry is a loathsome collection of theme parks and dead values', a thesis that was expanded upon at great length in Robert Hewison's *The Heritage Industry* (1987) and Patrick Wright's *On Living in an Old Country* (1985). This was all far removed from the vision of Octavia Hill and Hugh Dalton.

This is important because there is a clear pattern here and one that I think is highly relevant to our deliberations today at this conference. Although the push for change has often come from the elite vanguard, time and time again the driver has been the need to address social change in periods of rapid economic and cultural change. They responded to what the public wanted, and what society needed. Our predecessors both in and outside government were the radicals of their day.

Amidst today's globalisation and the challenge of building a multiracial, multicultural society, can we re-create that sense of a heritage movement, and one that is right for 21st-century Britain? I want us to learn from the success of those in the 'green' movement, which has become a worldwide force to be reckoned with by instilling a passion in communities. There is no doubt the popular will is there today. We see:

- 58 million visits to historic visitor attractions in 2004
- 3.4 million members of the National Trust and more than half a million members of English Heritage
- 157,000 volunteers within the sector
- 100,000 visitors to 310 sites in the first National Archaeology Week last July
- more than 3 million viewers to the BBC's *Restoration* programme and 2.5 million to Channel 4's *Time Team* programme.

To me it is no surprise that one of the heaviest uses of the internet is for genealogical family and local history research. In a society increasingly lacking the traditional social signifiers of class, religion and local labour markets, more and more people want to find out who they are, where they come from and what their roots are – even Jeremy Paxman!

Many organisations have risen to exploit this challenge. The Heritage Lottery Fund's Young Roots programme helps young people to find out about their heritage, using their interests and ideas, and their creativity and energy (Fig 37). These grants, each less than £50,000, show the Fund's commitment to funding innovative, community-focused projects across the country and to celebrating the country's 'hidden histories'.

We in government are being radical. We are reforming the way we designate and protect the nation's heritage. Many of you are involved. We are looking at how we can make the system more transparent, more open and more flexible. Perhaps most importantly, we are looking at how we give ownership back to the local communities themselves. We have some way to go. The heritage sector is perceived as experts talking to themselves. There is a lack of trust. The experts are seen as only willing to engage with communities on their own terms – and in a language that excludes those to whom they are talking.

If you do not want to believe the research, just listen to the experience of the people of Castleford. We will shortly be hearing the Castleford Choir singing about what heritage means to them. With support from both the Heritage Lottery Fund and English Heritage, this community has found its voice but it had to overcome a raft of obstacles first. And who put these obstacles in their way? It was us – the heritage community, the very people who claim to be representing what the public values. Castleford, you see, had been condemned: Pevsner called it a cultural wasteland; a Roman milestone was removed

Fig 38 A vanished national icon: Routemaster RM346 on its way to Streatham Station via Effra Road, Brixton, on the final day of its service on Route 159 in December 2005.
Guy-Howard-Evans, London Transport Museum © Transport For London

from the town to the British Museum on the grounds that the people of Castleford would not know how to look after it; and the Museums Service refused to bring objects out of store to show at the inaugural meeting on the grounds that they were priceless. The messages were clear: the heritage experts did not 'trust the community' or believe in what they were doing. Now, with a membership of more than 300, the Castleford Heritage Group has taken control and developed a new kind of relationship with the heritage experts. But perhaps the most interesting part of this story was what one of the heritage experts said himself:

> Much is made of the words 'facilitation', 'advocacy' and 'enabling' but no one had ever explained what these words mean. The Castleford project has given me an opportunity to explore what these words might mean – and by extension the skills required by heritage managers. Heritage management is about the technical aspect of conservation, but it is equally about encouraging and drawing out local skills, knowledge and experience of place rather than dictating what is of cultural significance.

Conceptions of heritage and 'whose heritage' are becoming all the more complex in modern Britain. How, today, do we nurture a national heritage? There is no easy answer – and, quite rightly, organisations like the Heritage Lottery Fund and the National Heritage Memorial Fund take a responsive approach to the issue. It is up to people, institutions and civil society to determine their own conceptions of heritage. Slowly, elements of national heritage – be they the National Health Service, Canterbury Cathedral, Hadrian's Wall, the Empire Windrush – accumulate within the national psyche.

We must all step up to that challenge. One way forward is the scheme I launched earlier this month – English Icons. Whether it is a cup of tea or Holbein's portrait of Henry VIII, when these symbols of Englishness are unpicked, they immediately reveal the multi-layered, multicultural element of English and British heritage. From the Romans to the Normans to the Dutch invasion of 1688, our island stories typically are global and then imperial. It was the tea-pickers of Sri Lanka and the Lascar dockers of Woolwich as much as the tea-merchants of Surrey who established the Englishness of a cup of tea.

Later in this conference you will look at your institutional values in the debate about 'Whose values matter?' I hope you will also ask yourselves again – as Tessa Jowell asked you in her essay last year – 'What more can we do to encourage greater diversity into both the heritage workforce and its audience?' We need to tackle this head on if we are to achieve the legitimacy to represent and advocate for the public value of heritage.

In Bradford, middle-class Asians are working hard to preserve their heritage – the mills from the industrial revolution. These buildings represent the beginning of their economic contribution to Britain. They help them feel part of Britain. Remember, too, that the Sikh community is travelling by bus up and down the country visiting sites and artefacts highlighted on the Anglo-Sikh heritage trail – a trail that highlights the contribution their community has made to Britain. And in London, where we said goodbye to the last Routemaster buses last month, the biggest tears were from the West Indian drivers and conductors for whom the Routemaster came to symbolise their journey to London in pursuit of a better life (Fig 38).

Our task is to revive that radical, empowering conception of heritage; to engage that mass of the public interested in the historic environment and its meaning for them; and to help build a Britain at ease with its present because it understands, values and is able to access its past.

Capturing the value of places – opening remarks
Sir Neil Cossons
Chairman, English Heritage

Until now, the debate about public value has been a closed conversation between 'experts' – which may be a euphemism for professionals – and politicians. But, of course, if values are to be real, the debate has to be much more open than this. We live in a world of changing perceptions. The advancing threshold of what the public sees as heritage means that the valued heritage gets larger. We value more but we throw away little. Quite properly, we never discard any of our older inheritances – the things and places that organisations like English Heritage and the National Trust have chosen to take into permanent captivity.

In English Heritage's case, the portfolio largely reflects 20th-century antiquarian values, which is part of the reason why several hundred independent museums and preservation groups were created in the 1970s and 1980s to cater for aspects of the heritage whose value was not then recognised by the big national institutions. The creation of civic societies in the same period was similarly a spontaneous desire by communities to protect buildings of under-appreciated local value.

As we meet to talk about public value, we need to remember that most of this country's listed buildings are privately owned, and that when we admire a historic town centre or a beautiful landscape from a railway train window, most of what we see – and value as part of the public good – is the private property of other people. So, in the historic environment, this debate about public value takes on a whole series of complex meanings.

When the historic environment sector came together to publish *Power of Place* (English Heritage 2000) there was a warm welcome for our vision of places as valuable, and equally strong agreement about the need for greater inclusivity and breadth of engagement in the way they are managed. However, some historic environment professionals were worried that their specialist role was about to be fatally undermined. In my view that is a false fear, provided that we are all prepared to accept a new contract with the public – that we recognise that our job is to care for and decode the past on their behalf. And, as we think about institutional values, we must remember too that people benefit from having their eyes opened by visionaries, madmen if you like, whose own knowledge and foresight enables them to see what others have yet to see, but whose enthusiasm enables them to open the eyes of a wider public. That is why we are celebrating this year the centenary of the birth of one of them – John Betjeman.

Let me finish with a few words about the Heritage Lottery Fund. As we all acknowledge with gratitude and admiration, the Heritage Lottery Fund has made an outstanding contribution to our national investment in the historic environment. Long may that continue. But, equally important, because it has taken the wider view of what the public values, heritage assets that would have been overlooked 10 or 20 years ago have been the beneficiaries of public money. That reflects changing public value. And the true beneficiaries are, of course, the public themselves.

One final point. It is especially important to this particular debate about public value: most of the public who will value what we do today have yet to be born.

Value and integrity in cultural and natural heritage: from Parks Canada to World Heritage

Christina Cameron
Canada Research Chair in Built Heritage, University of Montreal

Assessing the public value of heritage and putting historic values at the heart of decision-making are current subjects of interest in the heritage conservation world. My paper considers these questions from a practitioner's perspective. I will examine aspects of public value, including value attributed *to* heritage places, value derived *from* these places and values of the institutions responsible *for* them.

The title of my paper – 'Value and integrity in cultural and natural heritage' – uses vocabulary from the current Parks Canada system of managing heritage properties. Parks Canada has direct administrative responsibility for 42 National Parks, 2 National Marine Conservation Areas and 151 National Historic Sites; indirectly Parks Canada influences the management of some 700 other National Historic Sites – in other words, a vast and diverse array of natural and cultural heritage that covers more than 3 per cent of Canada's land mass.

The underlying theme of this conference raises the question of how to make the case for public investment in heritage conservation. This is linked to a perception that the public audience for heritage is a narrow segment of society, mainly middle class, and that there is a need for greater inclusiveness and a broader engagement of the public. It is also linked to a perception that heritage has been the exclusive domain of experts. The challenge set by the Secretary of State for Culture, Media and Sport, Tessa Jowell, in *Better Places to Live,* is that of finding new ways to talk about and engage others in the values and benefits of heritage.

Robert Hewison and John Holden of Demos have proposed a model for looking at public value, consisting of a triangle: intrinsic values of a heritage property; instrumental values like the social, economic and environmental benefits derived from a heritage property; and institutional values held by organisations responsible for heritage properties and programmes (see Fig 7). I have been asked to explore this triangle of values from a practitioner's perspective. How does this triangle of values actually translate into practice? To what degree has a heritage organisation like Parks Canada or indeed World Heritage created management systems taking into account this triangle of values? To what extent could a system that integrates all these values enhance inclusiveness and increase public participation in heritage matters?

'Intrinsic' values of heritage property

The use of the word 'intrinsic' is perhaps not the best one, since historic properties do not inherently have values. Historic properties take on value because people ascribe values to them. What makes a site part of our heritage is not the site itself but the fact that groups and individuals have attributed values to it. One can argue that all values are extrinsic, including physical ones. Values are complicated, multifaceted and diverse. A colleague of mine put it nicely: value 'lies in the associative and is expressed in the physical'. Or as Hamlet remarked: 'There is nothing either good or bad but thinking makes it so.'

People have been attributing value to historic places and natural areas for centuries. Indeed, most governments have well-established designation systems. They would certainly argue that significant heritage sites have always been managed on the basis of their values. It has been an informal system that made a lot of assumptions about significance. More recently, heritage conservationists have realised the importance of formally recording historic values and sharing them with others, as a necessary prerequisite to appropriate management. Australia was the first country to consciously articulate and record the heritage and cultural values of properties. Other countries have also been moving towards formal written statements of value or significance, as an essential tool in making decisions about a site.

The 2005 Operational Guidelines of the World Heritage Committee call for a formal statement of Outstanding Universal Value. The committee inscribes sites on the World Heritage List if they meet one or more of the ten criteria used to determine Outstanding Universal Value. In the past, the committee inscribed sites using the criteria but rarely completed the process by requiring a formal articulation and recording of their values. Canada discovered this last year when we prepared our Periodic Report for the committee on our 13 existing World Heritage Sites. Most of them lacked a formal statement of World Heritage values.

A weakness of designation systems in general, and World Heritage in particular, is that they segment values into various parts. The World Heritage Committee, for example, focuses only on Outstanding Universal Value. It does not concern itself with other values attributed to a site. This can create management problems for the site, especially if the local population does not understand or share those values. One need only think of the wanton destruction of the Bamiyan buddhas in Afghanistan to appreciate the dangers of a lack of connection between global values and local values.

Fig 39 Grosse-Île and the Irish Memorial National Historic Site, Canada. © *Parks Canada Agency*

An interesting example of this type is Tongariro National Park in New Zealand, a site that became World Heritage in 1990 because of its outstanding natural values (active and extinct volcanoes, and its mountain landscapes). How to explain this attribution of values to the Maori people who have lived in and cared for Tongariro for centuries and who attach spiritual and cultural significance to the mountains? Indeed, the Maori people gifted the park to the people of New Zealand in 1887. In this particular case, the government came back to the committee to ask for reconsideration of the site as a cultural landscape of Outstanding Universal Value, thereby recognising its cultural values.

The partitioning of values remains a concern for countries that have to look after World Heritage Sites. The issue is still under discussion, as witnessed by an international meeting of experts in Amsterdam in 2003. The conference topic was entitled 'Linking Universal and Local Values: Managing a Sustainable Future for World Heritage' (UNESCO 2004). A diverse group of experts from all continents and several scholarly disciplines presented case studies and engaged in a lively debate about the need to consider community values in the management of World Heritage Sites.

Fig 40 Irish cross at Grosse-Île and the Irish Memorial National Historic Site. © Parks Canada Agency

Canada has had similar experiences with regard to designations of national significance. Although more than 80 per cent of proposals for national designation come from the Canadian public, the actual selection is done through an expert advisory committee known as the Historic Sites and Monuments Board of Canada. The board takes a broad view of significance. But it was not always so. In the early 1950s, a national commission on the arts chided the board's preoccupation with military history ('historic sites are not just battlefields') and urged it to reflect more broadly on Canada's national life and common achievements. Board criteria now state that, to be designated, a place must have had a nationally significant impact on Canadian history, or must illustrate a nationally important aspect of Canadian human history. While one can argue that such criteria are necessary to determine national significance, one must none the less admit that this process sometimes fails to engage local interests, which attribute other values to the site. Indeed, in rare instances, the process infuriates special interest groups.

An example that may resonate with this audience is the site known as Grosse-Île, an island in the St Lawrence River downstream from Quebec City (Fig 39). The British government created a medical inspection and quarantine station on the island in 1832 and passengers immigrating to Canada were examined there until it closed in 1937. As part of a deliberation on the theme of Canadian immigration, the Historic Sites and Monuments Board selected Grosse-Île as the place to commemorate two aspects of the theme: the importance of immigration to Canada, particularly through Quebec City, from the early 19th century until the First World War, and the island's role as a quarantine station for the port of Quebec.

What happened next is a matter of public record. A chorus of complaint arose from Irish–Canadian organisations and individuals who felt passionately that 'their' Grosse-Île had not been adequately recognised in the general immigration story. What they meant by 'their' Grosse-Île was the tragedy of the typhus epidemic that played out on this island in 1847, in which some 5,000 people, mostly from Ireland, died in the space of one cruel summer (Fig 40). In response, government carried out public hearings in several Canadian cities, from the Atlantic to the Pacific, with a special focus on Ontario and Quebec. At the end of the process, the board proposed adding a third value to the site – the tragic experience of Irish immigrants, especially due to the 1847 typhus epidemic.

In addition, the site was renamed as Grosse-Île and the Irish Memorial National Historic Site of Canada.

Through experiences such as this, Parks Canada developed a policy approach that provides an opportunity to include all values that people attribute to sites. In its 1994 *Cultural Resource Management Policy*, provision is made for the evaluation of resources to determine which ones are cultural resources and what constitutes their value. The system allows for two levels of cultural resources, those of national significance being Level 1 and those having historical, aesthetic or environmental qualities as determined by other processes, including public input, being Level 2. This approach has been described by others as innovative, since few heritage management systems have such an inclusive approach to determining historic value. Its strength derives from three fundamental aspects: respect for the layers of history and all historic values, incorporation of built fabric and meaning (tangible and intangible) and encouragement of public participation.

Level 1 and Level 2 values refer to values attributed by people to the heritage property. Put simply, values refer to the reasons why people care. They are determined through formal designation processes but also through consultation with communities and stakeholders. In the Parks Canada system, all the values associated with the site are formally set down in a Commemorative Integrity Statement.

In the *Cultural Resource Management Policy*, values are linked to the concept of integrity (Fig 41). That is the 'integrity' part of this paper. Integrity is not a value. It refers to the state of wholeness or health of a historic place. In other words, it is the wholeness or health of its values. In the case of natural heritage, Parks Canada began using the concept of ecological integrity for its National Parks programme in the 1980s. The historic sites programme then adapted the concept to its needs, developing the notion of commemorative integrity.

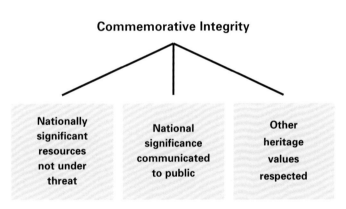

Fig 41 The concept of commemorative integrity in Parks Canada's Cultural Resource Management Policy.

The policy definition reads as follows: 'A historic place is said to possess commemorative integrity when the resources that symbolise or represent its importance are not impaired or under threat, when the reasons for its significance are effectively communicated to the public and when the heritage value of the place is respected.' The first element is straightforward – conservation of nationally significant resources. The second element underlines the need to communicate significance as a means of engaging and educating the public. The third element speaks to the issue of respecting all the values, including local values.

What this means in practice is that the Parks Canada system puts historic and cultural values at the heart of its decision-making processes. It also means that an empty dialogue pitting one value against another is defused. The approach is holistic and inclusive, requiring public participation and making a place for the management of community values alongside the national values.

Other benefits

The second set of values in the Demos triangle includes a whole range of benefits that are derived from a heritage property, including economic, environmental and social benefits. Civil servants and heritage groups have honed their skills to a fine point, crafting arguments and justifications to encourage investment in heritage because of its role in supporting other government or community priorities. Indeed there is an emerging corpus of literature on measuring such benefits.

Proponents argue that investment in historic places brings economic benefits through tourism, employment and sustainable development. Revitalised downtown areas stimulate business development. Property values in heritage districts typically increase, as do tax revenues. Heritage areas are tourism destinations of choice, witnessed as we have seen by a marked increase in tourism at World Heritage Sites. Parks Canada estimates that its system of parks and sites contributes more than a billion dollars to the Gross Domestic Product. Investment in heritage creates employment, since the renovation sector tends to be labour intensive and employs twice the number of people per dollar spent, in comparison with new construction.

Determining the economic benefits of heritage is a thriving sub-industry. Australian professor David Throsby has proposed a system for the economic evaluation of heritage, examining direct and indirect values as well as introducing the useful concept of 'cultural capital' (Throsby 2001). In France, Xavier Greffe also proposes economic models to determine the economic value of heritage (Greffe 2003). American professor Randall Mason has produced several research studies on the value of heritage for the Getty Conservation Institute (eg Avrami, Mason and de la Torre 2000). Two American economists also make the case. Donovan Rykema sees heritage properties as a differentiated product that commands a monetary premium (Rypkema 2003). Storm Cunningham argues that natural and cultural heritage offers economic opportunities in the trillion-dollar range as global industries restore natural and human environments when new land for development runs out (Cunningham 2002).

Investment in heritage also makes sense from an environmental perspective. Some argue that reusing existing buildings consumes 27 per cent less energy than new construction, thereby capitalising on the embedded energy invested in the original structures. By remaining within existing urban footprints, historic districts avoid new burdens on the water, sewer and transport systems, thereby contributing to sustainability. There is also a reduction in the amount of waste going to landfill. In Canada, recent estimates suggest that construction and demolition waste accounts for 20 per cent of materials going to clogged landfill sites.

In addition to economic and environmental values, there clearly are social values associated with heritage conservation. While social values are more difficult to determine, they are usually linked to identity and a sense of connection. Public opinion polls usually indicate high support for history and historic places, although methodologies differ and findings are sometimes inconclusive.

No matter how well the case is made for the economic, environmental and social value of historic places, getting political attention has generally been difficult. In Canada, most of these arguments fell on deaf ears at Treasury Board and Finance, until the Auditor General undertook the first-ever review of cultural heritage. Among her findings was the shocking statement that two-thirds of the built heritage managed by the federal government was in fair to poor condition, and that the same situation appeared to prevail for the National Historic Sites outside government ownership. The Auditor General made a passionate appeal to government, pointing out that 'Once our heritage is lost, it is lost forever!' This story ran in the national media for almost two months, probably because it was a national story with local roots. Each community had its own example of a government-owned building or canal that was being neglected. As a result of this sustained media attention, government committed substantial new funding.

Institutional values

The third point of the triangle covers heritage organisations themselves. What kind of institutional values will contribute to inclusiveness and a broader engagement of stakeholders in heritage matters? What behaviours and actions can the public expect of a heritage institution? What role does the public have in shaping institutional values?

These questions are rarely considered. Does the World Heritage organisation exist only for World Heritage Sites or is it about broader global heritage efforts in capacity-building and education? The same question arises for Parks Canada. Over the years, Parks Canada has debated whether its mandate should focus on its heritage places ('we are what we operate') or whether the organisation itself should demonstrate broad national leadership in the field. The answer, of course, is both.

Institutions normally have mandate statements, usually focused on the heritage properties, not on institutional values. Take English Heritage, for example. It reads as follows: 'English Heritage exists to protect and promote England's spectacular historic environment and ensure that its past is researched and understood.' Or Parks Canada: 'On behalf of the people of Canada, we protect and present nationally significant examples of Canada's natural and cultural heritage, and foster public understanding, appreciation and enjoyment in ways that ensure their ecological and commemorative integrity for present and future generations.'

Yet organisations rarely set down the way they will behave in carrying out their mandate. Parks Canada went through a crisis in organisational culture in the late 1980s and early 1990s, following years of budget cuts and government attempts at privatisation. The public no longer knew what to expect. Staff surveys and external polling confirmed that Parks Canada was confused about its own corporate identity and was sending out mixed messages to its various audiences. Some employees believed that the principal institutional value was business savvy (ie behaving like the private sector); others held fervently to the view that it was conservation; still others believed that it was contribution to the environment; and still others thought that the principal institutional value was to support tourism and economic development.

Following a period of institutional confusion and departmental change, Parks Canada emerged in 1998 as an independent government body with a clear mandate. One of the unusual features of the new agency was the legislative requirement for the chief executive officer to 'establish a charter for the agency that sets out the values and principles governing the provision of services by the agency to the public and the management of human resources of the agency'. The charter, then, was both externally and internally focused. It was meant to capture at a high level the way the agency would behave, what the public and employees could expect from the agency, and how the agency would be held accountable.

Without a clear roadmap, Parks Canada developed the charter through broad consultation internally and externally. And the results had to fit on one page! As an example of how elaborate the development process became, the internal dimension – a statement of values and principles – was created by a union-management team and involved two rounds of consultation with more than 5,000 employees. A consensus view settled on three human-resource-management principles: competence, fairness and respect. The external values were also subject to lengthy consultation. The final version of the charter, adopted in 2002, set out Parks Canada's role in deceptively simple language: guardians, guides, partners and storytellers.

Despite that simplicity, the Parks Canada Charter captures in a formal way a significant shift in the way the institution carries out its work. Of course, changes in corporate culture occur gradually. The charter simply documented a profound change in corporate behaviour that had been developing at Parks Canada over a number of years. It can be characterised as a shift from independent control by experts and officials to engagement and inclusiveness. I offer a few examples to demonstrate how the agency lives its charter values.

We are guardians of the national parks, the national historic sites and the national marine conservation areas of Canada.

In its role as guardian, Parks Canada traditionally would have limited its work to the sites that it administers directly. Now, the organisation has redefined its role to one of broader national leadership. In simple terms, it means that the organisation now shares its expertise with others involved in heritage conservation. It also means that Parks Canada supports the 'family' of National Historic Sites in third-party ownership, that it leads the development of national tools like the Canadian Register of Historic Places and the Standards and Guidelines for the Conservation of Historic Places in Canada, and that it advocates against inappropriate development proposals.

We are guides to visitors from around the world, opening doors to places of discovery and learning, reflection and recreation.

In the past, Parks Canada looked at visitors as a pressure to be 'managed'. Indeed, the organisation had a complex process called VAMP (Visitor Activity Management Planning) to identify what visitors could and could not do at heritage properties, as determined by the institution. This approach clearly does not put the visitor first! Parks Canada is now shifting gears to put a new focus on visitor experience. In the spirit of the charter (opening doors), the desires and expectations of potential visitors are now central to rethinking the range of services and opportunities on offer, with a view to facilitating memorable visitor experiences.

We are partners, building on the rich traditions of our aboriginal people, the strength of our diverse cultures and commitments to the international community.

Parks Canada automatically reaches out to partners and stakeholders in carrying out its work. The agency legislation itself requires a biennial gathering of a Round Table of interested Canadians to advise the minister on the performance of Parks Canada. The Round Table is in lieu of a proposed board of directors, opposed by stakeholders during Parliamentary debate as potentially too exclusive. The successful engagement of provinces and territories in the Historic Places Initiative has also been lauded as a model of a new way of working within our federal structure. Other partnerships include special relationships with stakeholders like the cruise-ship industry, tourism associations, chambers of commerce and ethno-cultural groups. Perhaps the single greatest achievement in positive relationships is the Aboriginal Advisory Committee, a group of indigenous chiefs and elders who regularly meet with the chief executive officer to discuss issues of mutual concern.

We are storytellers, recounting the history of our land and our people – the stories of Canada.

Heritage places have the power to connect us with our past in a visceral way. They are the physical embodiment of the creation of life and our human stories. Recognising the power of storytelling, Parks Canada invests in a range of communications approaches to reach audiences effectively. A national curriculum strategy aims at influencing schools to include natural and cultural heritage studies as part of the learning of every student. School visits to parks and sites are encouraged. Innovation funding has paved the way for new interpretative techniques and a greater diversity of voices. There is not just one story. There are many stories to be told.

In the 2005 corporate plan, the chief executive officer expressed strong commitment to inclusiveness and engagement:

> The only truly effective way for Parks Canada to fulfil this mandate is to build long-term, effective partnerships with Canadians. The more that we at Parks Canada work in harmony with as many Canadians as possible, the more successful we will all be in building a culture of heritage conservation in Canada.

Conclusion

To capture the public value of heritage is complex. There are many prisms through which heritage needs to be viewed, some easier to quantify than others. Each offers an opportunity to reach out beyond traditional audiences to engage others in what is important work: identifying and managing heritage for present and future generations. It is a work of creativity and consultation. It is forward-looking. It is an engagement with society. The key is greater inclusiveness, a broader engagement of the public and support from the public. This is the path forward to make the case for public investment in heritage.

Why do places matter? The new English Heritage *Conservation Principles*

Edward Impey
Director of Research and Standards, English Heritage

This conference has been about a search for connections: connections between the needs of modern society and the contribution that can be made to its well-being by the buildings, the places and indeed the whole environment – inevitably 'historic' – that we have inherited from the past.

My aim in this paper is to help move the discussion from the high ground of the broad social and economic 'instrumental' potential of heritage to a closer look at what the new language of values means for the management of the historic environment itself.

In particular I will be focusing on a framework of historic environment values that English Heritage is developing as part of its programme to set out a new set of clear and unambiguous principles for the sustainable management of the historic environment.

Identifying and classifying historic environment values (that is to say the types of value or 'meaning' or 'importance' which people may attach to elements of the historic environment) will, we believe, enable a much clearer understanding of what matters to people about the places around them.

This in turn should lead to better decisions about how aspects of the historic environment to which values are attached can be sustained (ie preserved and maintained) for the benefit of people today and in the future.

But first, it is worth reminding ourselves that the idea of value is far from a new one as far as the heritage professions are concerned. The value of old and beautiful places is what motivated the 19th-century founders of the Society for the Protection of Ancient Buildings and the National Trust. It was also a belief in the value of the past for the present that energised the champions of the first Ancient Monuments Act in 1882.

What has changed very profoundly is the range of subject matter to which we now attach value. In those early days, the officially recognised heritage was confined to the burial mounds and the castles that interested the learned Fellows of the Society of Antiquaries. But since then we have come to recognise that heritage is not confined to individual monuments or particular places but lies within – and even composes – the whole environment around us. One of the main reasons for this broadening of what we recognise as heritage is of course the huge energy that has been invested over the past 50 years by archaeologists and historians, amongst other specialists, in investigating and unravelling the story of our historic environment. And expertise will always have an absolutely vital role in the management of the historic environment and the attribution of value.

At the same time, the constituency of people attaching value to the historic environment has now broadened to include everyone. 'Ordinary people' – so described by Dame Liz Forgan, the Chair of the Heritage Lottery Fund, with genuine respect in her introduction to one of this conference's sessions – have become much more conscious of the legacy of the past that is all around them, and in turn their opinions and perceptions deserve to be listened to with much greater attention than was once the case.

It was in response to this inadequately recognised popular concern with the past that five years ago English Heritage and others published *Power of Place* (English Heritage 2000). At its heart was the acknowledgement of two basic truths.

The first is that the past, present and future cannot be separated, but form an inextricably linked continuum. The business of conservation is thus not about preserving historically significant places on their own, frozen at some particular time, but allowing them to coexist in sustainable harmony with an ever-changing present.

The second truth is that historic places do not have just one immutable value, but many overlapping values that reflect differing viewpoints. These are liable to evolve along with changes in people's own perceptions and interests, although longstanding attachment of value to places itself confers a species of value and adds substance to the idea of 'established' value referred to previously by the Secretary of State (see p 11).

It is therefore useful to identify and distinguish between types of value, and to do this we have used a version of the Demos triangle of heritage values (see Fig 7).

At one level there are the 'consequential' or 'instrumental' values that allow valued places to make a beneficial contribution to society as a whole – and these of course make up much of the subject matter of the earlier sections of these conference proceedings:

- educational   *a resource for learning*
- recreational   *a place for enjoyment*
- economic   *an asset for growth*
- social   *a force for cohesion*

Underpinning these 'practical' benefits there are the 'primary' or 'intrinsic' values of a place that together, but in varying combinations and proportions, give it its special significance:

- evidential   *people having access to the facts*
- historical   *people connecting with the past*
- aesthetic   *people visually responding to places*
- community   *people associating themselves with places*

And finally there are different kinds of 'institutional' or 'behavioural' values that those professionally involved in heritage need to respect in our role as mediators between places and the people who may want to make changes to them:

- communicating   *explaining historic value to others*
- listening   *hearing the perceptions of people*
- mediating   *negotiating solutions between values*

Fig 42 Fountains Abbey from Anne Boleyn's seat in Studley Royal Water Garden. The landscaped park and the ruins of the Cistercian abbey became a World Heritage Site in 1986 on account of their outstanding historical and aesthetic value.
© *NTPL/Andrew Butler*

My main interest in this paper is with the middle one of these families: the one that is about the 'value' of places themselves.

Just to clarify what I mean, I am going to offer some examples of places to which the particular values within this family might most readily be attached. For example Fountains Abbey and the landscaped grounds created around are appreciated for their 'aesthetic' value (Fig 42) while the Albert Memorial (see Fig 34) has been valued for 150 years as a great monument to our national identity and thus for 'historical' reasons. 'Evidential' value (ie what we can or may learn from something) is of course offered by both Fountains and the Memorial, but much more exclusively by those elements of the heritage that can

Fig 43 Archaeologists excavating Oxford Castle. Sometimes it is only through the expertise of specialists that the story of the buried past can be unravelled and made accessible to the public. © Oxford Archaeology

only be understood through specialised archaeological processes that probe what is otherwise hidden beneath turf, tarmac or even the surface of the sea (Fig 43). By and large this type of value is easily understood only by a few with special interests or training, and its meaning has to be interpreted and transmitted by them to the rest of the world. The role of the expert (whether a paid professional or a passionate amateur) as an agent or intermediary between the asset and the public is thus of vital importance.

And then at a fourth and much more extensive level there are the more everyday legacies of the past that add 'community' or 'personal' value to people's lives. These are the inherited structures and spaces, whether large or small, which add character and distinctiveness to our day-to-day surroundings; sometimes they may also be valuable for their deeper historical and aesthetic significance, but often simply for their familiarity and contribution to a sense of place (Fig 44).

The idea that decisions about changes to historic places need to take account of the opinions of the general public as well as the advice of experts was first championed by pioneering non-governmental organisations such as Common Ground, but has since found its way into the mainstream of heritage management. The Burra Charter, published by the Australian Committee of the International Council on Monuments and Sites (ICOMOS 1988), was the first major document to explain how the conservation of heritage sites has to be informed by a clear understanding of their significance: why is this place important, and for whom? And having answered those questions we need then to ask which aspects of it do we need to preserve or sustain and what can we safely allow to change and evolve?

Since then the principles of values-based conservation have begun to take firm root in the UK and across the world. In the process, however, it has become clear that many of our traditional approaches to measuring the value of historic places are too narrow and simplistic for the needs of today's complex plural society. What we have also learnt is that the dearly held heritage values of one person, or body, or group, can all too easily find themselves in conflict with the equally valid but different values of others.

Fig 44 A K2 telephone kiosk in the entrance to Burlington House, Piccadilly, London. The much-loved K2 came into service in 1923 following a competition won by Sir Giles Gilbert Scott. By the end of the 1930s more than 30,000 red telephone boxes had become quintessential elements of the British street scene.
Images of England © Stephen Hodgson

As we look ahead to the government's reform of the heritage designation system and the publication of a new Planning Policy Statement on the historic environment, we are therefore in urgent need of a much clearer set of ground-rules about what we mean by the value of heritage, and how that value can in turn best be sustained in the face of competing demands for change.

In response, English Heritage launched a consultation on a set of high-level *Conservation Principles* for use by English Heritage staff and others in the sector (English Heritage 2006). In parallel, the Institute of Field Archaeologists and the Institute of Historic Buildings Conservation are planning to adopt a comparable 'values-led' system of professional standards and guidance to the one we are proposing for ourselves at English Heritage.

The English Heritage principles may be summarised as follows

What is it that is valued – and who decides?
1 The historic environment is a shared resource
2 The values of the historic environment should be sustained
3 Everyone can make a contribution

Which places are significant – and why?
4 It is vital to understand the heritage value of places

How is change managed – and by whom?
5 Places should be managed to sustain their significance
6 Decisions about change should be reasonable, transparent and consistent
7 It is essential to document and learn from decisions

The first purpose of the principles is to establish a much clearer set of widely agreed definitions of what we all mean by 'heritage' and its value:

- what does it encompass, and how does it relate to the natural heritage?
- who is it for, and what benefits can it bring?
- why is there a legitimate public interest in the future of private historic assets and what is the balance necessary to recognise the contribution owners make to the maintenance and enhancement of the historic environment?

Secondly, the principles set out to provide a more powerful and inclusive way of finding out where that value lies:

- which are the places that we think are significant?
- to whom do they matter? Is it just to the experts, or to a much wider community of interested parties?
- do we value them because of their historic and aesthetic interest alone, or because of their economic value, or simply because people feel personally attached to them?

Thirdly, the principles are all about laying the path to a more transparent way of making decisions about how the value of historic places should or should not be sustained:

- what can change and how can the value of places best be conserved?
- how are the differing interests of conservers and developers to be reconciled?
- what values do the heritage organisations need to adopt to make this work?

Ensuring that changes in the historic environment are made in ways that sustain the values of a place in the most appropriate and effective way is a central aim of the *Conservation Principles*. How that happens depends entirely on the characteristics of the place itself – its original function, its present condition and the range of options on offer for its future use.

At one end of the spectrum, the rare and fragile values of a country house such as the National Trust's recently acquired Tyntesfield can now only be sustained by preserving the building as a non-functional visitor attraction and educational resource.

In other circumstances, the most effective way of conserving the values of a place is to allow it to continue in its original use, albeit in a discreetly updated and modernised form, as for example through the refurbishment of the 19th-century mill-workers' houses in Nelson and other parts of northern England, which people are beginning to appreciate for their combination of practicality, character and community value (see Fig 15). Further down the line, there are some buildings whose values cannot survive unless they are allowed to adapt themselves to new and different uses (Fig 45).

Fig 45. In 2005, conversion of the Great Central Warehouse into a library for the University of Lincoln won a Royal Institution of Chartered Surveyors gold medal for conservation. © Lincoln City Council

There are many situations like these in which open discussion between owners, developers and heritage professionals can allow an appropriate balance to be struck between the interests of conservation and economic development. But of course there are many others in which the conflicts are far more difficult to resolve.

An example of such conflict, as yet only partially resolved, involves the construction of very tall buildings within existing historic city centres.

English Heritage and other conservation bodies have argued that their unmanaged proliferation in London could have a damaging effect on long-admired historic views of the city, regardless of the architectural merits of the individual structures. To help overcome this problem English Heritage and CABE published explicit guidance on the values-related issues that need to be taken into account when making decisions about very tall buildings (CABE/English Heritage 2003). The purpose of the guidance is not to prevent their construction altogether, but simply to make sure that cities and their skylines can evolve in ways that do not damage the pre-existing special values of the place.

A rather different example is that of Seahenge, the extraordinary Bronze Age timber circle that was suddenly exposed by the winter tides of 1999 on the north Norfolk coast (Fig 46). Here, the debate focused on the relative strengths of two kinds of established but incompatible values: on the one hand, the archaeological profession believed that Seahenge should be excavated and conserved off-site in order that its evidential and historical values could be sustained for the benefit and enjoyment of future generations. On the other hand, there were others who attached a spiritual value to the site and felt that it needed to be left to the mercy of the natural forces of the sea. In this instance, after a long and public debate, English Heritage decided that the evidential value – what could be learnt from it about the people and technology of the Bronze Age – should be the winner and that Seahenge should be saved from the waves. As vindication of the decision to preserve the timbers laboratory analysis subsequently proved that the circle could be dated with astonishing accuracy to the spring of 2049 BC.

A third example of the conflict of values is provided by the Dundas Aqueduct, which carries the Kennet and Avon canal over the River Avon near Bath. In this case John Rennie's superb late 18th-century Bath-stone creation had been repaired in the early 20th century in engineering brick. When further work was required more recently, a debate focused on whether the aesthetic values of the structure (ie as intended by the designer) or the evidential value of the unsightly repairs most deserved sustaining. In the end, after

Fig 46 In 1999 Seahenge was in danger of being washed away by the tides. This provoked a conflict of values. Should it be preserved for its scientific value, or as an ancient sacred place should it be left to the mercy of the sea? © English Heritage

much discussion, it was decided that the greater value was the aesthetic and as a result the latest round of repairs has sought to return the aqueduct to something approaching its original appearance.

So far so good. Solutions, for better or worse, can be found. But there are still some serious risks to be faced. In particular, we need to be careful not to be lulled into a false sense of security by our own rhetoric, and into thinking that by talking honestly and openly about values that everything will somehow be all right.

In the case of tall buildings, what happens if UNESCO, as the ultimate guardians of World Heritage Sites, chooses to take a more rigorous conservationist line than that proposed within the CABE/English Heritage guidelines?

And more locally how do we make sure that the voice of local communities is really heard when decisions are being made about the future of a cherished place? And moving from that very proper desire for greater inclusion, how do we ensure that we do not pitch ourselves into a chaotic confusion of voices in which we start to lose all sight of any firm grasp of the established value of places?

One answer is that we need to make greater use of conservation plans and statements for designated heritage sites and landscapes that contain an agreed statement of the overall 'significance' of a place – in other words the sum of the many different values it has for the various interested parties.

At a more modest, but potentially very powerful level, we also need more documents like Village Design Statements – locally produced reports that set out a community's own view of what matters and needs to be sustained through the local planning system.

So where do we need to go next?

In the first place we need to press on with the task of establishing a clear new understanding of what we mean by value in the historic environment.

In the second place we need to make further progress with the identification and capture of that value, whether it be for World Heritage Sites, for local Conservation Areas, or simply the undesignated landscapes and streets and villages which comprise the historic environment.

And thirdly we must build on the experience we have already gained in the use of conservation plans to provide developers and the managers of heritage with simpler and more transparent ways of reaching negotiated agreements. The aim of these must be to find best ways of sustaining the values of places, while still allowing them to play a positive part in the life of a fast-changing modern society.

In conclusion I would like to remind you of the conservation cycle that English Heritage sees as fundamental to effective management of the historic environment – a virtuous circle in which understanding, valuing and caring are the keys to unlocking the power of places. And unlocking it not only for our own enjoyment, but for that of future generations to whom we must attempt to hand on those places in as good or better a condition as that in which they were passed to us.

Capturing the opinions of people

Dame Liz Forgan
Chair, Heritage Lottery Fund

This morning has been pretty intense – we have been presented with some of the theory and ideas behind heritage and public value. Now we are going to be much more practical – we are going to find out what the public actually thinks. This is an issue that is hugely important for us at the Heritage Lottery Fund. We must never forget that our role is to give out money that comes from lottery players. Their views matter.

Heritage people tend to assume that it is slightly dangerous to ask people what they think. And to be fair, if you go up to someone cold in the street and ask if they think heritage is a priority you can get short shrift. But sit down and talk to people about what matters to them about the past and a very different picture emerges. People have a very sophisticated grasp of the issues that arise from looking after the heritage – if only we take the time to listen – and if the public is given a chance to talk to people who have made projects happen, some very positive and interesting views emerge.

There is a lot of talk at the moment about how to involve the public in decisions about lottery – and it tends to boil down to simply asking people to tick a box on the back of a ticket. That will never be a sensible way to go about complex choices or to evaluate projects costing thousands or millions of pounds.

What we are doing instead is consulting people at a strategic level. What should we keep from the past? What should be thrown away? Why? Who should choose? We have been doing some fairly intensive consultation with groups of people, giving them a chance to look at some projects in detail and to talk to the people who made those projects happen. If you have had a chance to look at *Our Heritage – Our Future*, the consultation document where we are asking people to help us think about our future, you will see that we are also thinking about setting up some user groups to help look at what we do. At the same time, we are working hard with our regional committees (chosen by public advertisement) to make sure that the people on them really are local to that area.

Today we have a chance to share with you some of the strategic public consultation we have been doing. Opinion Leader Research (OLR) have been helping us to run a series of Citizens' Juries, first in Slough and Glasgow, then with young people in the Thames Gateway, and more recently with people from the East Midlands and Wales. OLR brought groups of people together – literally off the electoral roll – in order to show them what we have been doing and to find out what they thought. Even more exciting, four of the people who took part in the Citizens' Juries have kindly given up their time to come here today to speak about their experiences (see p 92). Two of the people from the projects that they saw are also sitting with us in the audience.

The value of heritage – what does the public think?
Deborah Mattinson
Joint Chief Executive, Opinion Leader Research

Overview

This paper presents an overview of a recent piece of public engagement work conducted by Opinion Leader Research for the Heritage Lottery Fund, exploring the general public's views on the public value of heritage.

In summary, the participants in this consultation felt that heritage itself matters most because it adds to our knowledge, it gives us a sense of identity, it enables us to hand on what we value to future generations and it makes where we live more special and distinctive. Looking specifically at heritage projects, funding matters most when it delivers regeneration and economic growth, brings people and communities together and offers opportunities to develop individual learning, skills and confidence.

Objectives

The project was designed to enable members of the general public with no specific interest in heritage to consider the value of heritage from two main angles: why heritage matters (ie the intrinsic value of heritage) and the benefits of heritage projects (ie the instrumental values of heritage).

We also explored the institutional values and benefits that the public see the Heritage Lottery Fund contributing through its policy and practices, and the degree to which it is seen as a trusted organisation to distribute lottery money to worthwhile projects.

Public value and heritage

The public value citizen workshops, or 'Citizen's Juries' as they came to be known, were developed to correlate a values structure developed by the Heritage Lottery Fund and Demos (Fig 47) with lay perceptions of heritage value and the benefits of Heritage Lottery Fund funding. In effect, they were designed to test the Demos theory and to see if the public really do share these views of what constitutes the 'public value' of heritage, and if so, what kinds of language they use when discussing this topic.

The Demos approach to defining public value was used as a framework for the research design and interpretation but, critically, participants were allowed to develop their own thoughts spontaneously.

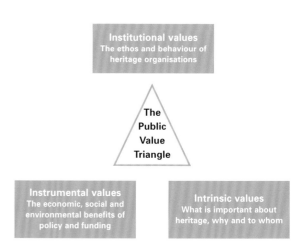

Fig 47 The public value triangle.

Our approach

We used a deliberative research methodology for this project, conducting two Citizens' Juries, one in the East Midlands (Nottingham, Derby) and one in Wales (Cardiff, Pontypool, Blaenavon). Each involved 16 participants, randomly recruited to fit the demographic profile of the area. We also ensured that the participants were not recruited to have a specific interest in heritage.

Over two and a half days, each jury reviewed, visited and heard testimony from five or six Heritage Lottery Fund-funded projects, covering a range of types and sizes of projects. Participants were then reconvened about a week later to identify the common themes over a further half-day session.

Deliberative approaches such as Citizens' Juries differ from standard market research in that they engage participants in an active dialogue over a longer period of time, presenting them with information, 'evidence' and, in this case, first-hand experience, to help them develop an informed point of view on an issue that they may not have considered in detail before. Independence and objectivity are also critical, delivered through independent recruitment methods to ensure that we do not just consult the 'usual suspects', and the involvement of Opinion Leader Research as an objective partner.

This family of research methods includes Citizens' Juries as well as other approaches such as Citizens' Forums and Citizens' Summits. We have been working with the Heritage Lottery Fund using deliberative research techniques since 2001 as they are a good way of helping the public to engage with a multifaceted issue such as heritage.

The projects

In each area, the Heritage Lottery Fund selected five or six projects for the jurors to review. The projects had all been successful in their grant applications, and were chosen as good examples of the wide range of projects funded by the Heritage Lottery Fund. Given that the aim of the research was to explore the public value of heritage, and lottery funding for heritage, it was important that all the projects were under way or complete, so it was not possible to include projects that were not successful in their application for a Heritage Lottery Fund grant.

East Midlands

In the East Midlands, the jurors heard from or visited five projects:

- the restoration and regeneration of Nottingham's Lace Market district, which houses the highest concentration of listed industrial buildings in England (Fig 48)
- the Nature Detectives, a Woodland Trust web-based project that encourages children to observe their natural environment and submit their information to build a national database; the website is accessible to all and there are specific materials for teachers and parents to use with children
- Black Text, a project that works with socially excluded young black people to help them research their roots and the history of famous black people, run by the Chase Action Group in Nottingham
- the West Shed near Derby, a locomotive shed, library and archive that focuses on Princess Royal class locomotives; its facilities have been significantly extended through a Heritage Lottery Fund grant
- Derby Arboretum, England's first public park, where the historic area has been returned to its original design, including the restoration of three listed buildings. The newer side of the arboretum has also been updated to include a new community building, recreation area and children's playground.

Fig 48 The restoration in 1996 of the Adams Building in Nottingham – a Grade II-listed former textile factory, lace warehouse and salesroom – was the result of a collaboration between the Lace Market Heritage Trust and New College Nottingham. It has since been cited by many as the starting point for the urban renewal of the Lace Market district.*
Ray Main © Heritage Lottery Fund

South Wales

In Wales, the jurors visited or heard from six projects:

- Pontypool Park, where work is being carried out to reinstate the park's historic character and original design, as well as to introduce new approaches to wildlife and grassland management
- St Peter's Church, a Grade II*-listed building in Blaenavon, where a Heritage Lottery Fund grant has been used for the provision of new heating and lighting systems, installation of facilities and repairs to the fabric of the building (Fig 49)
- Big Pit, the Gulbenkian-prize-winning national mining museum of Wales, where a Heritage Lottery Fund grant has been used to provide new exhibition space for the mining collections and to secure long-term access to the historic underground workings
- Cardiff Castle, where work is under way to repair the fabric of the castle itself, allowing public access to new areas, as well as to provide new visitor facilities and interpretation (Fig 50)
- On Common Ground, an activity project run by a team from the Museums and Galleries of Wales; this project works with young people across Wales to help them develop their own ideas about heritage through guided visits, artwork, films and oral history records
- a Prince's Trust project, in which a group of young people produced a DVD and play about the lives of people in Tredegar before the National Health Service and the impact and influence of Nye Bevan

Key findings

The jurors were asked to consider the public value of heritage from two angles: intrinsic value – or why heritage matters – and instrumental value – the benefits of heritage projects.

Intrinsic values

Looking first at intrinsic value, participants value heritage strongly and identified a wide range of intrinsic values of heritage, spontaneously covering almost all of the values identified through the Demos work. Four key priority values emerged:

Fig 49 St Peter's Church, a Grade II*-listed building in Blaenavon, where a Heritage Lottery Fund grant has been used for the provision of new heating and lighting systems, installation of facilities and repairs to the fabric of the building.
David Ward © Heritage Lottery Fund

- Knowledge value – which places heritage as central to learning about ourselves and society, understanding our cultural identities at both personal and community levels
    'The projects all matter for the same reasons; helping future generations to understand the past and how things have developed from the past.'
- Identity value – delivering a sense of identity on a personal, community, regional or national level – seen as an intrinsic value of heritage, as well as a benefit of project activities
    'Apart from being about heritage, the projects seem to be about community, identity, and the future and past.'
- Bequest value – heritage should be cared for in order to hand on things that are valued to future generations
    'It's nice to know how far we've come and if we didn't preserve it, we wouldn't know.' 'I think it's important to keep different types of heritage because they reflect a broad and multicultural society.'
- Distinctiveness value, or what makes somewhere special – a key spontaneous value for heritage, viewed as extremely important because it is closely linked to personal and cultural identity
    'It's important to keep an element of history in a rapidly modernising world in order to cement the area's character and historical meaning.'

Instrumental value

When considering the benefits of heritage projects – or instrumental benefits – participants again spontaneously identified a large number of the 'Demos' benefits as a result of their experience of Heritage Lottery Fund projects. Four key areas of benefits emerged:

- economic benefits – the top priority instrumental values of heritage projects overall are regeneration and economic growth. These benefits are seen as critical because they help to create a 'ripple effect' of well-being across a host of areas from employment to enjoyment and pride. [On Nottingham's Lace Market area]
    'The whole area has now been regenerated. It was a no-go-area with dilapidated buildings that has now been converted into the most exclusive area in Nottingham city centre.'

- the benefits to the area where a project is taking place – participants identified a whole range of benefits here, but the most important are the improved profile and reputation of an area, a safer and improved environment, reduction in antisocial behaviour and improved leisure opportunities. [On Derby Arboretum]

    'It has enabled the park to become a safe place that welcomes people of all ages and has something for them all.'

- benefits to the community affected by a project – while economic benefits obviously have a substantial impact, participants also identified softer values here such as greater public spirit, mutual understanding and pride in the local area, values which are distinctively heritage-related. This applies particularly to projects that celebrate and commemorate the history of ordinary people

    'The projects all unite communities with a reason to be proud of those that have gone before them.'

- benefits to individuals – benefits to individuals were also important, in particular learning, skills and confidence, which came through markedly in the review of smaller, activity-based projects, especially those involving young people. [About On Common Ground]

    'Young people who would otherwise see themselves as having no future discover new skills and interests and more about their background.'

Institutional benefits

While the Heritage Lottery Fund itself was not a key focus for the study there was a very positive response to the Heritage Lottery Fund as a result of information gained, and widespread agreement that the Heritage Lottery Fund was right to fund the projects considered.

> 'Because I've seen them first hand, things like this make me believe that they are good. If this is just a sample of what the Heritage Lottery does, then that's brilliant.'

Participants in Wales picked up on and praised the Heritage Lottery Fund's encouragement of, and support for projects by heritage funding 'cold spots'. Both juries were reassured by the Heritage Lottery Fund's approach to holding projects accountable and there was endorsement of the support that the Heritage Lottery Fund gives to projects before and after the application process.

There was also clear support for the Heritage Lottery Fund's broad view of heritage and notable support for intangible heritage projects, such as oral histories, as well as for the capacity to fund both large- and small-scale projects.

> 'It needs to be [a broad definition of heritage] because one person's definition of heritage is going to be different from another person's. And the Heritage Lottery Fund has to think about everybody.'

Other issues

One of the features of deliberative research is that participants often raise issues outside the focus of the consultation that turn out to be as illuminating as their views on the main issues.

A whole range of additional topics came up during this consultation, including:

- clear support for smaller, activity-based projects, which are valued for their tangible impact on participants and their sheer value for money

    'Amazing how much work has been done with £50,000!'

- the importance of inclusivity, which ties back to the key benefit of bringing communities together

    'The most important projects are the ones where they involve the whole community and a wide range of people can benefit from them.'

- a real and spontaneous concern about sustainability and support for the Heritage Lottery Fund's role in evaluating and monitoring projects

    'OK, the money's going in now but how sustainable is that, how long term is that?'

- praise for the enthusiasm and commitment of the project advocates and volunteers

    'What struck me is the sheer enthusiasm of the people who do these projects.'

- a desire to see projects promoted more widely to encourage use and appreciation

    'It makes you wonder how much the actual locals know and what history they've got on their doorstep.'

Impact of the process

Finally, the deliberative process had a clear positive impact on the participants in several ways. Their perceptions of the value of heritage shifted over the course of the event, in terms of both its personal and wider importance. Indeed, many more participants went away feeling that their own community had a rich heritage.

Some were also inspired to think about actively taking up heritage-related activities and there were even cases where attitudes had changed within a week – in Wales, for example, one participant revisited one of the projects (St Peter's Church in Blaenavon) in between the two stages and was planning another visit later in the month, and some initially sceptical Welsh participants had encouraged friends and family to visit Big Pit.

'I can see me going to the arboretum. I only live ten minutes away from it and I've never been before to see it, but this weekend, I'm definitely going.'

Participants also gained a more positive view of the Heritage Lottery Fund, feeling that it plays an important role in preserving heritage. And in some cases perceptions and the likelihood of playing the Lottery changed, with some participants who play regularly saying they would now feel more positive about how their money is being used, and a few non-players actively considering purchasing a ticket.

'I'm going to buy a lottery ticket. I've never done before. I think of it as dead money, but I'd consider buying a lottery ticket.'

The Citizens' Jury

Fig 50 A highlight of the conference was the discussion between Dame Liz Forgan and four members of the Citizens' Juries who have been visiting Heritage Lottery Fund projects around the country – Bunney Hayes and Dolly Tank from the East Midlands, and Michael Rosser and Leila Seton from Wales. Also contributing were two of the local managers who had been responsible for explaining their projects to the juries and listening to their feedback – Karen Aberg from the Derby Arboretum and Sarah Greenhalgh from the National Museums and Galleries of Wales.

Leila Seton: 'The Fund needs to do more to let people, and especially school-children, know what's happening locally. In my own case, I'd never been to the Big Pit mining museum and had no idea how amazing it is.'

Michael Rosser: 'Every project had something we liked, and we realised what an incredibly difficult job it is for the Fund to choose between them. My own favourite was Pontypool Park in Cardiff, which I didn't even know existed until we went there!'

Bunney Hayes: 'Citizen's Juries could certainly help with choosing projects, but only if they are given the information to really understand what they are all about. Sometimes it would also be useful to hear about the projects that don't make it through the system – and some projects are so important that they really ought to be funded directly through schools and not have to rely on the Lottery.'

Dolly Tank: 'I used to think heritage was just your own family history, not places like parks and museums. It was only when I joined the Citizens' Jury that I realised it is about things that are actually relevant to me personally. And when you buy a lottery ticket it's nice to see where the money's going!'

4 The Values of Heritage Institutions

Conference overview

Kate Clark
Deputy Director, Policy and Research, Heritage Lottery Fund

The final discussion session of the conference touched on the third type of value in the heritage version of the public value triangle, institutional values – the values shown by an organisation in how it operates. Trust, legitimacy and confidence are fundamental sources of public value. They can come from how an organisation behaves and what it delivers, but they are also rooted in general levels of trust within society. The ethos and culture of an organisation can determine the extent to which it can create and even destroy value. In schools, prisons and in the police force, the culture, climate and attitudes of staff and managers can have a direct impact on performance (Kelly *et al* 2002, 13).

But what are the appropriate values and behaviours for a heritage organisation? And are they any different to the values that any public sector organisation should display? Many different people touched on this, throughout the conference. As Tessa Jowell noted,

> The market place can tell us how many people visited a particular museum or how much profit a particular show or event made, ... but when it comes to putting a value on things like trust, fairness and accountability, it has failed miserably.

Her very clear message though, was that if heritage is to embrace public value, it

> will require a radically different mindset ... It means taking a genuine interest in what our citizens think, and not just consulting in a ritualistic and formulaic way because we have to. It means engaging a much wider swathe of society, particularly people who are socially excluded and people from ethnic minority communities, and not just talking to the usual suspects.

Parks Canada has already gone down the route of exploring its own institutional values. As Christina Cameron noted, although heritage organisations often have clear goals, they do not always explain how they will behave. Emerging from what was effectively a crisis of legitimacy in the late 1970s and 1980s, Parks Canada aimed to establish a charter for the agency that sets out its values and principles – looking both at how the agency should behave and what the public and employees could expect and how the agency would be accountable. The results fit on a side of paper and look deceptively straightforward, but the only truly effective way of fulfilling that mandate is to build long-term partnerships with Canadians, which in turn means a very different approach to heritage.

One of the real challenges that arose from their work centred round the role of experts in such partnerships. As the Work Foundation pointed out, a public value framework recasts the role of experts, using them to inform and empower the public rather than simply cutting the public out. But heritage experts are worried that their specialist role will be undermined. Sir Neil Cossons thought that this was

> ... a false fear, provided that we are all prepared to accept a new contract

with the public – that we recognise that our job is to care for and decode the past on their behalf.

He also reminded us of people such as John Betjeman who, as an expert, was also able to open people's eyes to the value of heritage. Indeed, the new *Conservation Principles* presented by Edward Impey stress this role (English Heritage 2006, 23):

> Practitioners should use their knowledge, skills and experience to encourage people to understand, value and care for their heritage. They play a crucial role in communicating and sustaining the established values of places, and in helping people to articulate the values they attach to places.

This is an issue that is hugely important for the Heritage Lottery Fund – Liz Forgan noted that

> We must never forget that our role is to give out money that comes from lottery players. Their views matter.

The problem is, she said, that heritage people tend to assume that it is slightly dangerous to ask people what they think. The Heritage Lottery Fund's own experience has been that experts can learn a lot by talking to the public, and at the same time – as the Citizens' Juries showed – people who are not necessarily heritage specialists can gain new insights into what heritage is all about. Indeed, the juries went to the heart of some of the key issues for the Heritage Lottery Fund – sustainability, involving communities and the importance of good project leaders.

Heritage champions are playing an increasingly important role in creating a voice for heritage specialists in local authorities; in the language of public value they help heritage people to engage with the many different stakeholders that make up their 'authorising environment'. Cllr Garnett spoke about her role in making practical connections between heritage and wider local authority agendas – communities, the economy and the very practical business of keeping North Yorkshire's hundreds of historic bridges open for use. By making these connections, heritage champions can engender trust by helping to overcome prejudices about heritage, which in turn can help local authorities see heritage as something widely beneficial.

It was David Lammy who reminded us of the complex relationship between specialists and society. When looking back at earlier heritage champions, he said that

> Although the push for change has often come from the elite vanguard, time and time again the driver has been the need to address social change in periods of rapid economic and cultural change. They responded to what the public wanted, and what society needed.

Yet the minister raised some tough challenges. He said that the heritage sector is perceived as experts talking to themselves, and identified a lack of trust. And he asked how we nurture a 'national heritage' in a world where concepts of heritage – and whose heritage – are becoming far more complex.

These questions became the core issues in the debate at the last session of the conference, which asked, 'Whose values matter?'

Discussion 3. Whose values matter?

Chair: Nick Higham
(Arts and Media Correspondent, BBC)

Panel

Bonnie Greer (British Museum Trustee)

Ben Rogers (Head of Democracy, Institute of Public Policy Research)

Simon Thurley (Chief Executive, English Heritage)

Graham Wynne, (Chief Executive, Royal Society for the Protection of Birds)

Professor the Baroness Lola Young of Hornsey

The Issues

- Do we get the heritage we deserve?
- Are heritage organisations in touch with what the public cares about?
- What happens when the views of experts and the public conflict?
- Do heritage bodies have the right organisational values to do their jobs properly?

The words below are not necessarily exactly those that were spoken, and are intended instead to capture the overall flavour of the discussion.

Chair: Over the past two days we have heard a lot about the ways in which the values of heritage can be measured, and we have been shown how the public is increasingly being drawn in to that process. However, the Secretary of State has also made it clear that as far as she is concerned there is still a need for a radically different mindset in the way our heritage institutions work and make their decisions. So, just how far are you, the professionals, in touch with the public, and what happens when you do not agree with one another?

Baroness Young

Things are getting more sophisticated, but the organisations themselves are not yet very diverse. Lots of people do not yet know how to make a connection between 'high heritage' and 'personal heritage'.

Chair: Are these legitimate criticisms?

Ben Rogers

There is hardly *any* institution that is *fully* in touch with the public. The public does, though, have a special appetite to get involved in heritage through television programmes (eg *Restoration*) and schemes like the Local Heritage Initiative – it is happening.

Chair: Are you working with an outmoded definition of heritage, and outdated ways of working?

Simon Thurley

The publication of *Power of Place* in 2000 opened our eyes to what exists beyond designation. English Heritage's role is to reflect the values held by everyone. This is quite easy in terms of classic heritage, but more difficult with non-core aspects of the heritage that people are only beginning to see as interesting and important. Conservation is actually much more democratic than ministers say, but it is true that the process has until recently been too dominated by experts. In future we are going to need a stronger two-way flow of ideas if we are to develop properly 'refined preferences'.

Ben Rogers

Simon Thurley is right: there are *some* unchanging core values.

Bonnie Greer

A new generation is coming along that has no conception of time and traditional heritage – for them Big Brother House is more important than Salisbury Cathedral. The British Museum is now asking the hard questions – like 'Why is this collection here in this country?' People are not being taught to see the value of heritage – without the eyes to see, there is no value.

Graham Wynn

The Royal Society for the Protection of Birds is actively trying to broaden its reach. In fact the RSPB was founded by some very feisty women but is now run by middle-class men. We know we need to do more about diversity in our organisation; we also need to reach out to more people and are doing that by, for example, opening reserves in the middle of urban areas. But at the end of the day, there is a multiplicity of present and future publics, and it is therefore incredibly difficult to find a single view of 'what the public wants'. Given the quagmire, some leadership is inevitable, even if it leads to charges of elitism.

Delegate

The government is moving on from pure economic value and starting to recognise social value and identity. To take this forward we now need a new paradigm that includes the contribution of psychology and moral philosophy to understanding of the past.

Bonnie Greer

Yes, we do have to go back to basics – and allow value to be bottom up, not top down.

Delegate

You talked earlier about a younger generation having no concept of heritage – are today's teenagers *really* that different from those of previous generations?

Bonnie Greer

Yes – our new digital world is changing *everything*. Fifty years from now people will probably not even go into museums, except for social interaction. We have no choice but to focus on the under-16s; the rest are a lost generation.

Ben Rogers

Adolescence now starts earlier and lasts much longer. Engagement with heritage is a huge challenge, given that young people will not even vote.

Chair: Do you agree that your audience is either dying or has lost interest?

Simon Thurley

There have been revolutions like this before; you just have to adapt. And it is important to remember that teenagers have never been focused in the past: it's not their thing.

Graham Wynn

The RSPB finds that the real interest is among children and middle-aged people, not the ones in between.

Delegate

There is real interest among young people when it comes to working with heritage. A survey by the Construction Industries Training Board recently showed that lots of them want to 'save old buildings' – and as evidence of that it is worth noting that all the college places for plasterers and stonemasons are now full.

Bonnie Greer

And don't forget the over-60s – they are also learning to declare their demands and interests.

Delegate

One inevitable aspect of a public-values-based system is that it will sometimes capture very fleeting values. Do we have the mechanisms for handling it?

Delegate

Do we not need to persuade the public that today's good design is tomorrow's heritage – that they are not separate things but a continuum?

Ben Rogers

The expert–public cleavage is not as bad as it is made out. If you give power to the public, they will when necessary kick the experts.

Chair: Who currently has the final decision on what's important, and would it be better if English Heritage was allowed to be the ultimate referee?

Simon Thurley

It is the democratic planning system that balances different perspectives. English Heritage has to take a view on behalf of the heritage, but should not be the final judge. Experts have a vital educating and mediating role in developing refined public preferences.

Delegate

The difference is that experts 'think' and 'know', whereas people 'feel' and 'believe'.

Delegate

We have to be aware of the danger of relativism in a world with no fixed boundaries. Do we need to make a clearer distinction between *personal* heritage and *permanent* heritage?

Baroness Young

Different bodies of knowledge reside in different publics; that is not necessarily the same as relativism.

Graham Wynn

But we need to acknowledge that *public* expertise is also important. And we do need to recognise that we operate in a vicious economic climate, which is why we have to have the public on our side.

Ben Rogers

We need to change the terms of exchange between ourselves and the public – or publics.

Bonnie Greer

My job as a trustee is to hand on something of value to the next generation, and at the same time to make the things we hold in trust relevant for living people. At the end of the day, it's about the people who come to our museums. And that's all that it has got to be about, because otherwise we are talking about dead things that don't mean anything to anybody.

Chair: Thank you to all our panellists, and to you the audience for your questions. And finally, may I offer our very special thanks to Kate Clark of the Heritage Lottery Fund. She's the one who put together the whole of this programme, so she's the one who has done all the work!

Overleaf: Fig 51 Since August 2001, Cardiff Castle has been undergoing a major programme of conservation. The £8 million project is being driven forward by Cardiff Council, with the support of a Heritage Lottery Fund grant of £5.7 million, the largest ever awarded in Wales. Part of the massive scheme is the creation of a new education centre, which is set to welcome its first school groups in September 2006. Mike Caldwell © Heritage Lottery Fund

Conclusion

'To be effective performance measures do have to garner the commitment of those whose performance they measure … if not, they risk demoralising and misdirecting rather than animating and inspiring those who are asked to do the work.' (Moore and Moore 2005, 91)

The warning is an apposite one. Yet the two days spent discussing public value suggest that this is not a risk. The quality of the papers, the noisy discussion during breaks, and the volume of mail that we have all received suggest that thinking about public value as an approach to heritage can be energising.

The critical factor that will mark the success of the event is what happens next. Each of the organisations involved is committed to taking forward public value in their own work.

The Heritage Lottery Fund is in the process of drafting its next strategic plan. The Fund wants to build on the Citizen's Juries and to increase the amount of public engagement in its work. It has already set out plans for doing so in the consultation on its next strategic plan. It will also be using public value as a framework for its social and economic research data, and to help review its own procedures and practices as part of its third strategic plan.

The Department for Culture, Media and Sport is exploring ways of creating a model of public value and how to apply this in future.

English Heritage has issued its new consultation document *Conservation Principles for the Sustainable Management of the Historic Environment*, which puts value at the heart of all conservation decisions.

The National Trust is publishing the results of its work with Accenture as *Demonstrating the Public Value of Heritage* and also considering the role of public value approaches in its own strategy review and operational management.

As we conclude, it is worth remembering these words spoken in 1889 by William Morris to the annual meeting of the Society for the Protection of Ancient Buildings:

> these old buildings do not belong to us only . . . they belonged to our forefathers and they will belong to our descendants unless we play them false. We are only trustees for those that come after us.

For those of us who follow in Morris's footsteps as public guardians of the heritage, there could be no more fitting reminder that our duty is not just to the places themselves, but to the people for whom they hold value, both today and in the future.

Ultimately the conference demonstrated that the heritage sector is committed to a more inclusive approach to heritage that recognises the real need to bring specialists and communities together. If public value is to be useful, it has to be more than a framework imposed from outside the sector with little understanding of what heritage is about. If it is to work, it has to enable us to share our passion for heritage with others, while at the same time helping us to engage with the realities of public service.

Heritage Lottery Fund; English Heritage;
Department for Culture, Media and Sport; National Trust

Biographies of speakers

Accenture
Accenture is a global management-consulting, technology-services and outsourcing company. Committed to delivering innovation and with deep business-process expertise and broad global resources, Accenture helps its clients to become high-performance businesses and governments. With more than 126,000 people in 48 countries, the company generated net revenues of US$15.55 billion for the fiscal year ended 31 August 2005.

Baroness Andrews
(Elizabeth) Kay Andrews, Baroness Andrews of Southover, was appointed Parliamentary Under-Secretary of State at the Office of the Deputy Prime Minister in May 2005. She was previously a Government Whip and Spokesperson for Health, Work and Pensions, and for Education and Skills. Before being raised to the peerage, Baroness Andrews was a Fellow of the Science Policy Research Unit at Sussex University, Parliamentary Clerk in the House of Commons, and Policy Adviser to Neil Kinnock as Leader of the Opposition.

Ricardo Blaug
Dr Ricardo Blaug is Senior Lecturer in Political Theory at the University of Leeds. He is an award-winning author who has published more than 30 articles and books on democracy, participation and organisational change. Formerly a psychiatric social worker, he is now a director of Research Republic, a public policy research consultancy, and an Associate Consultant at The Work Foundation.

Dr Christina Cameron
As Director General of National Historic Sites, Christina Cameron provided national direction for Canada's historic places, focusing on heritage conservation and education programmes. Since 2005, she has occupied the Canada Research Chair in Built Heritage at the University of Montreal. She has participated in World Heritage work since the mid-1980s and has written numerous articles on the evolution of the Convention, its strengths and weaknesses, and the challenges that countries face in conserving World Heritage Sites.

Kate Clark
Kate Clark is Deputy Director, Policy and Research, at the Heritage Lottery Fund. She is an archaeologist, specialising in industrial archaeology. She has worked with the Ironbridge Gorge Museum, the Council for British Archaeology and English Heritage. Her main interests are in how people value heritage and in developing techniques for heritage facilitation. She is the author of the Fund's guidance on conservation planning, as well as books and articles on values, building recording and industrial archaeology.

Sir Neil Cossons
Neil Cossons became Chairman of English Heritage on 1 April 2000. A leading authority on the history of technology and industrial archaeology, he previously directed the Science Museum, the National Maritime Museum, Greenwich, and the Ironbridge Gorge Museum in Shropshire. A past President of the Museums Association, he was the first Chairman of the Association of Independent Museums and is currently its President.

Dame Liz Forgan
Liz Forgan joined the National Heritage Memorial Fund and Heritage Lottery Fund as Chair in April 2001. Following an early career in journalism, she moved to television with the start of Channel 4 where she became Director of Programmes. In 1993 she joined the BBC as Managing Director of BBC Radio. She is a former Chair of the Churches Conservation Trust and also a former Trustee of the Phoenix Trust.

Heather Garnett
Heather Garnett began her career as a sheep and beef farmer. Next, she set up an enterprise offering training in counselling. In the 1990s, when her children left home, she entered university following a first degree with an MA in architectural history. In 2001 she was elected to North Yorkshire County Council where, in addition to her role as Heritage Champion, she developed her understanding of local government through membership and chairmanship of several committees.

Robert Hewison
Robert Hewison has written widely on 19th- and 20th-century British cultural history. He was Slade Professor of Fine Art at Oxford University, taught English Literature at Lancaster University, and is a regular contributor to *The Sunday Times*. He is an Associate of the independent think-tank Demos. His recent books include *Culture and Consensus: England, Art and Politics Since 1940* (1997), *Towards 2010* (2000), *Ruskin's Venice* (2000) and, with John Holden, *The Right to Art* (2004) and *Challenge and Change* (2005).

John Holden
John Holden is Head of Culture at Demos. His main professional interest is in the development of people and organisations in the cultural sector. He

has been involved in numerous major projects in this field for the Department for Culture, Media and Sport, the Museums, Libraries and Archives Council, the Clore Foundation and other bodies. He was a principal organiser of the influential Valuing Culture conference in June 2003.

Louise Horner

Louise Horner is Associate Director at The Work Foundation and leads the organisation's work on public value. Her numerous reports include studies of public-sector pay, industrial relations, performance in the public and private sectors, and the future of work. She was previously a policy analyst at the Strategy Unit in the Cabinet Office, where she developed policy on leadership in the public sector and on workforce development.

Edward Impey

Edward Impey has been Director of Research and Standards at English Heritage since November 2002. He was appointed Assistant Curator, Historic Buildings with the Historic Royal Palaces Agency in 1995 and was later Curator of the same institution. He is a member of the Institute of Field Archaeologists and a Fellow of the Society of Antiquaries. His personal research interests include the medieval architecture of England and Normandy.

Rt Hon Tessa Jowell, MP

Tessa Jowell was appointed Secretary of State for Culture, Media and Sport in 2001 and Minister for Women in 2005. She has been the MP for Dulwich and West Norwood since 1992 and is a visiting Fellow at Nuffield College, Oxford. She was previously Minister of State for Employment, Welfare to Work and Equal Opportunities at the Department for Education, and Minister of State for Public Health at the Department of Health. Before her election to Parliament, Tessa had a career in psychiatric social work, social policy and public-sector management.

David Lammy, MP

David Lammy was appointed Parliamentary Under-Secretary of State at the Department for Culture, Media and Sport in May 2005. He was previously Parliamentary Under-Secretary at the Department for Constitutional Affairs and Parliamentary Under-Secretary for Health. Before entering Parliament, David was a member of the Greater London Authority with a portfolio for culture and arts. He is a trained lawyer and was called to the Bar in 1995.

Rohit Lekhi

Rohit Lekhi is the founder of Research Republic, a public policy research consultancy, and an Associate Consultant at The Work Foundation. Formerly an award-winning teacher of politics at the University of Warwick, he is a highly skilled researcher and has written numerous articles, books and reports on social mobility, race and ethnicity, public-service reform and market regulation.

Deborah Mattinson

Deborah Mattinson is one of Britain's leading practitioners of issue-based research and consultation. In 1992, she co-founded Opinion Leader Research, of which she is joint Chief Executive. Deborah writes and broadcasts widely on public opinion, citizen engagement, corporate social responsibility and political polling. She is also a co-founder of the Smart Company, a consultancy specialising in corporate social responsibility, and is a Commissioner of the Equal Opportunities Commission and a Trustee of the Green Alliance.

Julia Thrift

Julia Thrift joined the Commission for Architecture and the Built Environment (CABE) in June 2003 as the founding Director of CABE Space. Before joining CABE Julia spent five years at the Civic Trust, the national charity that campaigns for improvements to the built environment. Prior to this she spent 10 years as a journalist, writing about design for a wide range of national newspapers and specialist journals. She is a Fellow of the Royal Society for Arts.

David Throsby

David Throsby is Professor of Economics at Macquarie University in Sydney, Australia. He has published widely in the economics of the arts and culture, and the relationship between cultural and economic policy. His research has focused on demand and supply in the performing arts, the role of artists as economic agents, culture in economic development, sustainability and cultural capital, and the economics of heritage conservation. He is a past President of the Association for Cultural Economics. His most recent book, *Economics and Culture*, was published 2001.

Sue Wilkinson

Sue Wilkinson is Director of Learning and Access at The Museums, Libraries and Archives Council (MLA). Her remit also covers responsibility for the Council's regional agencies and for its flagship museums project Renaissance in the Regions. Sue has been responsible for developing Inspiring Learning for All, which was launched in March 2004. Before joining MLA she was Deputy Education Officer at HM Tower of London and Director of the South Eastern Museums Education Unit.

References

Avrami, Erica, Mason, Randall and de la Torre, Marta 2000. *Values and Heritage Conservation*. Los Angeles: The Getty Conservation Institute

CABE 2002. *The Value of Good Design: How Buildings and Spaces Create Economic and Social Value*. London: Commission for Architecture and the Built Environment

CABE 2003. *Building Sustainable Communities: Actions for Housing Market Renewal*. London: Commission for Architecture and the Built Environment

CABE 2004a. *The Value of Public Space: How High Quality Parks and Public Spaces Create Economic, Social and Environmental Value*. London: Commission for Architecture and the Built Environment

CABE 2004b. *Is the Grass Greener? Learning from International Innovations in Urban Green Space Management*. London: Commission for Architecture and the Built Environment

CABE 2004c. *Manifesto for Better Public Spaces*. London: Commission for Architecture and the Built Environment

CABE 2005a. *Does Money Grow on Trees?* London: Commission for Architecture and the Built Environment

CABE 2005b. *Parks and Squares: Who Cares?* London: Commission for Architecture and the Built Environment

CABE 2005c. *Physical Capital: How Great Places Boost Public Value*. London: Commission for Architecture and the Built Environment

CABE/English Heritage 2003. *Guidance on Tall Buildings*. London: Commission for Architecture and the Built Environment and English Heritage

Carpenter, Evelyn 2004. *Out of the Hopeless Box. Creative Neighbourhoods: An Evaluation*. London: Arts Council

Clark, Kate (ed) 1999. *Conservation Plans in Action: Proceedings of the Oxford Conference*. London: English Heritage

Cunningham, Storm 2002. *The Restoration Economy: The Greatest New Growth Frontier*. San Francisco: Berrett-Koehler

DCMS 2006. *Better Places to Live: Government, Identity and the Public Value of Heritage. Summary of Responses*. London: Department for Culture, Media and Sport

de la Torre, Marta, MacLean, Margaret, Mason, Randall and Myers, David 2005. *Heritage Values in Site Management: Four Case Studies*. Los Angeles: Getty Conservation Institute

Demos 2004. *Challenge and Change: Heritage Lottery Fund and Cultural Value*. London: Heritage Lottery Fund and Demos

DoE/DNH 1994. *Planning Policy Guidance Note 15. Planning and the Historic Environment*. London: Department for the Environment and Department of National Heritage

English Heritage 2000. *Power of Place: The Future of the Historic Environment*. London: English Heritage for the Historic Environment Steering Group

English Heritage 2005. *Heritage Counts 2005: The State of England's Historic Environment*. London: English Heritage

English Heritage 2006. *Conservation Principles, Polices and Guidance for the Sustainable Management of the Historic Environment*. London: English Heritage

Fishkin, James S 1992. *The Dialogue of Justice: Toward a Self-Reflective Society*. New Haven: Yale University Press

Greffe, Xavier 2003. *La Valorisation économique du patrimoine*. Paris: La Documentation française

Hamilton, Alexander, Madison, James and Jay, John 1961. *The Federalist Papers*. New York: Mentor Books

Heritage Lottery Fund 2005. *Conservation Management Plans: Helping your Application*. London: Heritage Lottery Fund

Hewison, Robert 1987. *The Heritage Industry: Britain in a Climate of Decline*. London: Methuen

Hutter, Michael and Rizzo, Ilde (eds) 1997. *Economic Perspectives on Cultural Heritage*. London: Macmillan

ICOMOS 1988 (revised 1999). *Australia ICOMOS Charter for the Conservation and Restoration of Monuments and Sites*. Canberra: Australia ICOMOS

Jowell, Tessa 2004. *Government and the Value of Culture*. London: Department for Culture, Media and Sport

Jowell, Tessa 2005. *Better Places to Live: Government, Identity and the Value of the Historic and Built Environment*. London: Department for Culture, Media and Sport

Kelly, Gavin, Mulgan, Geoff and Muers, Stephen 2002. *Creating Public Value: An Analytical Framework for Public Service Reform*. London: Strategy Unit, Cabinet Office

Matarasso, François 1997. *Use or Ornament? The Social Impact of Participation in the Arts*. Stroud: Comedia

Moore, Mark 1995. *Creating Public Value: Strategic Management in Government*. Cambridge (Mass): Harvard University Press

Moore, Mark and Moore, Gaylen Williams 2005. *Creating Public Value Through State Arts Agencies*. Minneapolis: Arts Midwest and The Wallace Foundation

Navrud, Ståle and Ready, Richard C (eds) 2002. *Valuing Cultural Heritage: Applying Environmental Valuation Techniques to Historic Buildings, Monuments and Artifacts*. Cheltenham: Edward Elgar

Peacock, Alan (ed) 1998. *Does the Past Have a Future? The Political Economy of Heritage*. London: Institute of Economic Affairs

Rypkema, Donovan 2003. 'Economic benefits of heritage conservation'. Keynote address at Canadian Association of Municipal Administrators, Winnipeg, 27 May, pp 29–44.

Throsby, David 2001. *Economics and Culture*. Cambridge: Cambridge University Press

UNESCO 2004. *Linking Universal and Local Values: Managing a Sustainable Future for World Heritage*. World Heritage Series 13. UNESCO: World Heritage Centre

Wright, Patrick 1985. *On Living in an Old Country: The National Past in Contemporary Britain*. London: Verso

Delegates attending the conference

Karin Aberg	Derby Arboretum
Andrew Allen	National Trust
Jan Ambrose	Building Conservation
Christian Andersen	Fonden Realdania
Judith Anderson	Historic Scotland
Nadine Andrews	Arts About Manchester
Wendy Andrews	Wendy Andrews Public Relations
Baroness Andrews	Office of the Deputy Prime Minister
Josie Appleton	Spiked
Sam Ashby	Office of the Deputy Prime Minister
Chris Atkins	Department for Culture, Media and Sport
Louise Austin	Cambria Archaeology
Mary Austin	Heritage Lottery Fund
Brian Ayers	Norfolk Museums and Archaeology Service
Andrew Backhouse	Culture North West
Kevin Baird	Heritage Lottery Fund
Janet Barber	Kaleidoscope Research and Policy
Gordon Barclay	Historic Scotland
Helen Barnard	Royal Society for the Protection of Birds
Paul Barnwell	English Heritage
Constance Barrett	Churches Conservation Trust
Dave Barrett	Derbyshire County Council
David Barrie	Channel 4
Mandy Barrie	Department for Culture, Media and Sport
David Barrie	National Art Collections Fund
Susie Barson	English Heritage
Marion Barter	Architectural History Practice
Mark Bates	Department for Culture, Media and Sport
Liz Bates	Heritage Trust of Lincolnshire
Hilary Beal	Multimedia Ventures
Owen Bedwin	Essex County Council
Nancy Bell	National Archives
Harriet Bell	Student
Robert Belli	Church Commissioners
Peter Bembridge	Heritage Open Days
Richard Paul Benjamin	University of Liverpool
Simon Bennett	Museums, Libraries and Archives West Midlands
Jodie Bettis	Student
David Bevan	Bedfordshire County Council
John Bevan	Church Commissioners
Robert Bewley	English Heritage
Liz Bingham	Countryside Agency
Ricardo Blaug	The Work Foundation
Geoffrey Bond	Archives, Libraries and Museums London
David Bonnett	David Bonnett Associates
Sue Bowers	Heritage Lottery Fund
Carol Bowsher	West Midlands Hub
Steve Brace	Royal Geographical Society
Joyce Bridges	English Heritage
Matthew Brigden	Accenture (UK) Ltd
Nick Brigland	Historic Scotland
Jessica Britton	Local Heritage Initiative
Richard Brookes	*Sunday Times*
Karen Brookfield	Heritage Lottery Fund
Andrew Brown	English Heritage
Sarah Brown	English Heritage
Adrian Browning	Church Commissioners
Adrian Budge	Royal Armouries Museum
Catherine Bunting	Arts Council England
Henry Burgess	Department for Culture, Media and Sport
Neil Burton	Architectural History Practice
Tony Burton	National Trust
Elaine Cabuts	National Museum Wales
Jane Callaghan	Multimedia Ventures
Amanda Callard	Woodland Trust
Judith Calver	Green Places
Christina Cameron	University of Montreal
Emily Candler	National Museum Directors' Conference
John Carman	University of Birmingham
Emma Carver	English Heritage
Jay Carver	Scott Wilson
Amanda Chadburn	English Heritage
Chris Chandler	UK Film Council
Jill Channer	The Prince's Regeneration Trust
Helen Charlton	Sussex Arts Marketing
Gill Chitty	Council for British Archaeology
Kate Clark	Heritage Lottery Fund
Andy Clark	Purcell Miller Tritton LLP
Roger Clark	YHA (England and Wales) Ltd
Sian Clarke	Department for Culture, Media and Sport
Adam Clarke	Heritage Education Trust
Madeleine Clegg	Department for Culture, Media and Sport
Judith Cligman	Heritage Lottery Fund
Nigel Clubb	English Heritage
Ralph Cobham	Resource Consultants International
Sue Cole	English Heritage
Katya Condy	Heritage Open Days
Jenny Cooper	British Waterways
James Cooper	Woodland Trust
Helen Corkery	Arts About Manchester
Sir Neil Cossons	English Heritage
Ben Cowell	Department for Culture, Media and Sport
Jamie Cowling	Department for Culture, Media and Sport
Lisa Cox	Heritage Lottery Fund
Alexandra Coxen	English Heritage
Karl Creaser	English Heritage
Stephen Creigh-Tyte	Department for Culture, Media and Sport
Dorian Crone	English Heritage
Debbie Dance	Oxford Preservation Trust
Prakash Daswani	Cultural Co-Operation
Alan Davey	Department for Culture, Media and Sport
Maurice Davies	Museums Association
Sue Davies	Wessex Archaeology
Stephen Davis	Woodchester Mansion Trust
Andrew Davison	English Heritage
Fiona Davison	London Museums Hub
Geoff Dawe	English Heritage
Michael Dawson	Surrey County Council
Samantha Dawson	TiC Consultants
Tom Dawson	University of St Andrews
Anne Dixon	National Trust
Nick Dodd	Sheffield Galleries and Museums Trust
James Doeser	Institute of Archaeology
Torin Douglas	BBC Social Affairs Unit
Alison Drake	Castleford Heritage Trust
Paul Drury	Paul Drury Partnership
Richard Dumville	English Heritage
Ian Dungavell	Victorian Society
Rosemary Elder	Chartered Institute of Building (CIOB)
Birgitta Elfstrom	Swedish National Heritage Board
Sally Embree	English Heritage
Keith Emerick	English Heritage
Richard Evans	Creative and Cultural Skills
Martin Fairley	Historic Scotland
Kate Fellows	Historic Houses Association
Anna Ferguson	English Heritage

John Fidler	English Heritage	Edward Impey	English Heritage
Jon Finch	Museums, Libraries and Archives North West	Robert Isherwood	University of Manchester
Gareth Fitzpatrick	Heritage Education Trust	Jane Jackson	Edinburgh World Heritage
Brendan Flanagan	Cheshire County Council	Laura Jackson	National Art Collections Fund
Tony Fleming	English Heritage	Dianne Jefferson	Department for Culture, Media and Sport
Dame Liz Forgan	Heritage Lottery Fund	Tiffany Jenkins	*The Scotsman*
Katie Foster	HLF West Midlands Committee	Soegaard Jensen	Bygningskulturelt Rad
Kate Frame	Historic Royal Palaces	Stephen Johnson	Heritage Lottery Fund
Jenny Freeman	Historic Chapels Trust	Daniel Johnston	Jacobs Babtie
Ylva French	Campaign for Museums	Christine Johnstone	Inland Waterways Amenity Advisory Council
Mark Friend	BBC	Sam Jones	Demos
David Gaimster	Society of Antiquaries	Helen Jones	Victoria and Albert Museum
Andy Ganf	Archives, Libraries and Museums London	Rt Hon Tessa Jowell, MP	Department for Culture, Media and Sport
Paul Gardner	Queen's College, Oxford	Barry Joyce	Derbyshire County Council
Rita Gardner	Royal Geographical Society	Jamie Kaminski	Brighton Business School
Steve Garland	Bolton Museums and Art Gallery	Suvi Kankainen	National Museum of Science and Industry
Heather Garnett	North Yorkshire County Council	Anna Keay	English Heritage
Frances Garnham	Historic Houses Association	Carleen Keleman	Cornwall County Council
Pedro Gaspar	War Memorials Trust	Beatrice Kelly	Heritage Council of Ireland
Abigail Gilmore	Culture North West	Claudia Kenyatta	Department for Culture, Media and Sport
Gwen Gittens	Department for Culture, Media and Sport	Lesley-Anne Kerr	CyMAL, Museums Archives and Libraries Wales
Sharon Goddard	Heritage Lottery Fund	Sharon Kerr	Gateway Gardens Trust
Dawn Goodfellow	Devonshire Educational Trust	Sarah King	Association of English Cathedrals
Jennifer Gosling	Ancoats Buildings Preservation Trust	Sally King	Countryside Agency
Mary Beth Gouthro	Student	Mary King	King Partnership
Catherine Graham-Harrison	Heritage Lottery Fund & National Heritage Memorial Fund	Naomi Kinghorn	University of Newcastle
		Karen Knight	Student
Christopher Gray	National Maritime Museum	Victoria Lamb	Department for Culture, Media and Sport
Sarah Greenhalgh	National Museums and Galleries of Wales	Deborah Lamb	English Heritage
Bonnie Greer	British Museum	Alison Lammas	Countryside Agency
Jane Grenville	University of York	David Lammy, MP	Department for Culture, Media and Sport
Neil Grieve	University of Dundee	Louise Lane	Heritage Lottery Fund
Doreen Grove	Historic Scotland	Patricia Langley	Heritage Lottery Fund
Camilla Hampshire	Royal Albert Memorial Museum, Exeter	Jennifer Latto	Heritage Lottery Fund
Lord Donald Hankey	Gilmore Hankey Kirke Ltd	Beverley Lear	Lear Associates
Jan Trane Hansen	Kulturarusstyrelsen	Rebecca Lee	Renaissance Yorkshire
Stephen Harding	Royal Institute of British Architects	Rohit Lekhi	The Work Foundation
Richard Harris	Essex County Council	Richard Lesley	Local Heritage Initiative
Andrew Harris	Scott Wilson	Miriam Levin	English Heritage
Richard Harris	Weald and Downland Open Air Museum	David Levy	BBC
Geoff Harrison	PLB Consulting Ltd	Viola Lewis	Museums, Libraries and Archives
Ian Hart	National Audit Office	Robyn Llewellyn	Heritage Lottery Fund
Imelda Havers	BlueFish Regeneration Ltd	Jeremy Lucas	Essex County Council
Will Hawkesworth	Colchester Borough Council	Ian Lush	Architectural Heritage Fund
Bunney Hayes	Citizens' Jury	Greg Luton	English Heritage
Janice Hayes	Warrington Library, Museum Archives Service	Fiona Lydon	Arts Council England
Brian Hayton	Hull Museum and Art Gallery	Fiona MacDonald	Berkshire Archaeology
David Heath	English Heritage	Alistair Macdonald	Urban Practitioners
Martyn Heighton	National Historic Ships	Helen Maclagan	Warwickshire County Council
James Hervey-Bathurst	Historic Houses Association	Frances MacLeod	Department for Culture, Media and Sport
Robert Heslip	Student	Gareth Maeer	Heritage Lottery Fund
Robert Hewison	Demos	Ekta Malhotra	Accenture
Ben Heyes	Bolton Abbey Estate	Patricia Mandeville	Department for Culture, Media and Sport
David Hicks	Edinburgh World Heritage	Peter Martin	Essex County Council
Nick Higham	BBC News	Thomas Martinsen	Bygningskulturelt Rad
Lisa Hill	Arts and Humanities Research Council	Randall Mason	University of Pennsylvania
Jennifer Hill	Student	Amanda Mathews	Countryside Agency
Guy Hills-Spedding	Ministry of Defence	Deborah Mattinson	Opinion Leader Research
John Hoadly	National Audit Office	Nicholas Mayhew	Ashmolean Museum
John Holden	Demos	Duncan McCallum	Policy and Communications Group
Louise Horner	The Work Foundation	Alastair McCapra	Institute of Conservation
Sandi Howie	Northern Ireland Environment & Heritage Service	Sarah McCarthy	Prince Research Consultants Ltd
Sue Howley	Museums, Libraries and Archives Council	Greg McErlean	Royal Parks
Sophie Hunt	Hampshire County Council	Julie McGuinness	National Audit Office
Tristram Hunt	Heritage Lottery Fund	Paul McLaughlin	Historic Royal Palaces
Gregor Hutcheon	National Trust	Rita McLean	Birmingham Museums and Art Gallery

Colin McLean	Heritage Lottery Fund	Ingrid Samuel	Department for Culture, Media and Sport
Jim McLelland	Sustain	Mhora Samuel	Theatres Trust
Anna McPherson	Paul Drury Partnership	Toby Sargent	Department for Culture, Media and Sport
Ian McQuiston	Historic Buildings Council	Kathryn Sather	Kathryn Sather and Associates
Victor Middleton,	Tourism Society	Paul Schurer	English Heritage
Sue Millar	Sue Millar Associates	Alison Scott	Heritage Lottery Fund
Sophia Mirchandani	South East Museum, Library and Archive Council	David Sekers	Heritage Lottery Fund
Helen Monger	Heritage Lottery Fund	Simon Seligman	Devonshire Educational Trust
John Moore	Historic Monuments Council, Northern Ireland	John Sell, CBE	Joint Committee of National Amenity Societies
Louis Moreno	Department for Culture, Media and Sport	Leila Seton	Citizens' Jury
Katharine Morley	National Coal Mining Museum for England	Gretta Sharkey	Castleford Heritage Trust
Jane Morris	*The Guardian*	Bob Sharpe	South West Museums, Libraries & Archives Council
Gerri Morris	Hargreaves McIntyre		
Olivia Morris	National Trust	Eleanor Sier	Student
Rebecca Morris	Opinion Leader Research	Ross Simmonds	English Heritage
Michael Morrison	Purcell Miller Tritton LLP	Faye Simpson	Museum of London
Andrea Mulkeen	Church Commissioners	Simon Smales	North Yorkshire County Council
Vivek Nanda	Alan Baxter and Associates	Jeremy Smith	Community Channel
Amanda Nevill	British Film Institute	Robert Smith	Heritage Lottery Fund
Palmer Newbould	Former Trustee of National Heritage Memorial Fund and Heritage Lottery Fund	George S Smith	Southeast Archaeological Centre, Tallahassee
		Chris Smith	English Heritage
Andrew Newman	University of Newcastle	David Smurthwaite	National Army Museum
Antonia Nichol	National Trust	Chris Smyth	Heritage Railway Association
Laura Norris	Alba Conservation Trust	Kathleen Soriano	National Portrait Gallery
Tim O'Connor	Talkback Productions	Carole Souter	Heritage Lottery Fund
Sean O'Reilly	Institute of Historic Building Conservation	Fiona Spiers	National Heritage Memorial Fund
Adrian Olivier	English Heritage	Mary Sabina Stacey	Combe Down Stone Mines Project
Lucy Ord	Museums, Libraries and Archives North West	Sarah Staniforth	National Trust
Liz Owen	Opinion Leader Research	David Steele	UK Film Council
Ece Ozdemiroglu	Economics for the Environment Consultancy Ltd	James Stevens	English Heritage
Stephen Page	National Trust	Heather Stewart	British Film Institute
Helen Palmer	Palmer Squared	Sheila Stone	Heritage Lottery Fund
Georgia Parks	Department for Culture, Media and Sport	Anthony Streeten	English Heritage
John Paton	Hornet Dynamics	Frank Strolenberg	Projectleider Belvedere
Joshua Peck	Heritage Lottery Fund	Anna Strongman	Arup
Garrett Peters	Crown Estate	Rosslyn Stuart	English Heritage
Adrian Philips	National Trust	Alan Sutherland	Department for Culture, Media and Sport
Caroline Pick	East Midlands Museums, Libraries and Archives Council	Hedley Swain	Museum of London
		Bob Sydes	North Yorkshire County Council
Victoria Pirie	Creative and Cultural Skills	June Taboroff	Gilmore Hankey Kirke Ltd
Mike Pitts	*British Archaeology*	Fred Taggart	The Prince's Regeneration Trust
Ian Poole	English Historic Towns Forum	Eddie Tait	Historic Scotland
Colette Price	Countryside Council for Wales	Fiona Talbott	London Borough of Hackney
Edwina Proudfoot	Scottish Church Heritage Research	John Tallantyre	Department for Culture, Media and Sport
Lee Pugalis	One North East	Virginia Tandy	Manchester City Galleries
Kate Pugh	Heritage Link	Dolly Tank	Citizens' Jury
Layla Pyke	Strategy and Projects Regeneration	Matthew Tanner	*SS Great Britain*
Louise Ray	National Council on Archives	Andy Tatman	Department for Culture, Media and Sport
Mark Reay	Royal Society for the Protection of Birds	David Thackray	National Trust
Simon Rees	Student	Emyr Thomas	Countryside Council for Wales
Harry Reeves	Department for Culture, Media and Sport	Rachel Thomas	English Nature
Lucy Regan	Heritage Lottery Fund	Katrina Thomson	Clore Fellow 2005–6
Fiona Reynolds	National Trust	Julia Thrift	CABE Space
David Rhodes	Independent heritage adviser	David Throsby	Macquarie University, Sydney
Alan Richards	Welsh Historic Monuments	Simon Thurley	English Heritage
Paul Richardson	UK Film Council	Simon Timms	National Trust
Jacqueline Riding	Clore Leadership Programme	Steve Trow	English Heritage
Katie Roberts	Heritage Lottery Fund	Crispin Truman	Churches Conservation Trust
Terry Robinson	Countryside Agency	Peter Tullin	Arts and Business East
Peter Robinson	Institute for Public Policy Research	Janice Tullock	Clore Leadership Programme 2005–6
Annette Roe	Scott Wilson	Robin Turner	National Trust for Scotland
Ben Rogers	Institute for Public Policy Research	Michael Tutton	Society for the Protection of Ancient Buildings
Charlie Rose	Westminster City Council	Sjoerd van der Linde	Student
Nick Roslund	Department for Culture, Media and Sport	Pieter van der Merwe	National Maritime Museum
Michael Rosser	Citizens' Jury	Louise Vaux	Multimedia Ventures
Henry Russell	College of Estate Management	Laura Venn	Culture West Midlands

Philip Venning	Society for the Protection of Ancient Buildings	Laura Williams	South East Museum, Library and Archive Council
Becky Volker	GreenSpace		
Geoffrey Wainwright	Society of Antiquaries	Martin Williams	Tyne and Wear Museums
Christopher Walker	Arts and Humanities Research Council	Susan Williamson	Historic Scotland
Kathryn Walker Tubb	Institute of Archaeology, London	Kate Wilson	English Heritage
Nick Ware	Community Channel	Primrose Wilson	Heritage Lottery Fund
Catherine Ware	Heritage Lottery Fund	Ian Wilson	National Trust
Laura Warren	Department for Culture, Media and Sport	Chris Winter	English Historic Towns Forum
Anna Warrington	National Gallery	Fiona Wood	Department for Culture, Media and Sport
Giles Waterfield	Heritage Lottery Fund	Corinna Woodall	Heritage Lottery Fund
Anthony Waterman	Building Research Establishment Ltd	Rob Woodside	Atkins Heritage
Malcolm Watkins	Gloucester City Council	Kate Worsley	Civic Focus
Jonathan Watson	Lottery Monitor	Graham Wynne	Royal Society for the Protection of Birds
Margaret Way	Government Office for the West Midlands	John Yates	Institute of Historic Building Conservation
Emma Way	Landscape Design Trust	Janine Young	L-P Archaeology
Gill Webber	British Library	Baroness Lola Young of Hornsey	
Audrey Wedderburn	Countryside Agency		
Lizzie West	Department for Culture, Media and Sport		
Andy Wharton	Countryside Agency – Landscapes, Access and Recreation		
Rowan Whimster	Whimster Associates		
Jenifer White	English Heritage		
Richard Whittaker	English Heritage		
Greg Wilkinson	Accenture		
Jacky Wilkinson	Bath and North East Somerset		
Sue Wilkinson	Museums, Libraries and Archives Council		
Paula Williams	Countryside Council for Wales		

Exhibitors

Castleford Heritage Trust
Department for Culture, Media and Sport
English Heritage
Heritage Lottery Fund
Local Heritage Initiative
Market Leader Research
Museums, Libraries and Archives
Royal Society for the Protection of Birds

Fig 52 Archaeology for All: the Hidden Treasures of the Thames. Sorting historic artefacts found on the foreshore gives Londoners a hands-on opportunity to understand how people have interacted with the river throughout history. Partners in this Heritage Lottery project included Tate Modern, Museum of London, Thames21, Museum in Docklands, Greenwich Foundation, the Shadwell Basin Activities Centre and the Pumphouse Educational Museum.
© Heritage Lottery Fund